Threat versus Safety Theory (TVST)

A New Perspective on Disease, Disability, Pain, and Suffering

David R. Clawson

ISBN: 978-1-59849-379-5
Library of Congress Control Number: 2024923503

BISAC Codes:
HEA039000 HEALTH & FITNESS / Diseases & Conditions / General
SCI086000 SCIENCE / Life Sciences / General
SOC000000 SOCIAL SCIENCE / General

Printed in South Korea

Editors: Mir-Yashar Seyedbagheri and Danielle Harvey
Design: Soundview Design

www.threatversussafety.com

Classic Day Publishing
206-860-4900
info@classicdaypub.com
www.classicdaypub.com

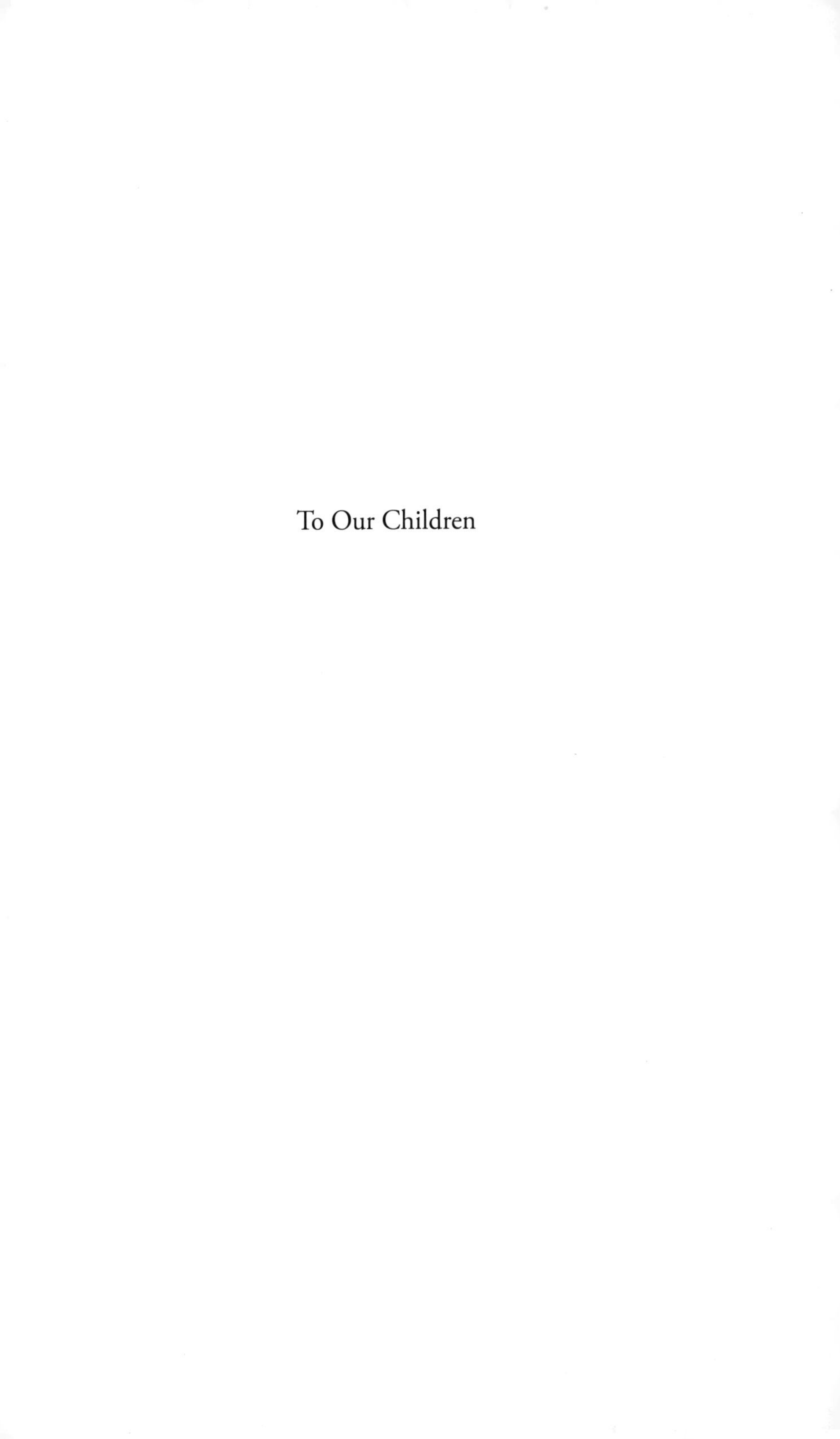

To Our Children

"The one who plants trees,
knowing that he will never sit in their shade,
has at least started to understand
the meaning of life."

Rabindranath Tagore

Thanks Amy, Bill, Bob, Danielle, Dave, David, Elliott, Garrett, Hara, Jake, Kim, Les, Linda, Mark, Mel, Mir-Yashar, Ray, Ruthie, Stephen, Steve, Sue, and the D.H.G.

"To perceive the world differently, we must be willing to change our belief system, let the past slip away, expand our sense of now, and dissolve the fear in our minds."

Gerald G. Jampolsky

Purpose – to protect and propagate the human genetic code.

Mission – to increase awareness of threat versus safety in disease, disability, pain, and suffering.

Vision – to create people, institutions, curricula, algorithms, policies, and laws committed to increasing safety within the world.

Introduction

"There is grandeur in this view of life, with its several powers, having been originally breathed into a few forms or into one; and that, whilst this planet has gone cycling on according to the fixed law of gravity, from so simple a beginning endless forms most beautiful and most wonderful have been, and are being, evolved."

Charles Darwin

Threat
/THret/noun/ a person or thing that is likely to cause danger, risk, injury, or illness.

Ver·sus
/vərsəs/preposition/ as opposed to; in contrast to.

Safe·ty
/sāftē/noun/ the condition of being protected from or unlikely to be caused danger, risk, injury, or illness.

The·o·ry
/THirē/noun/ a supposition or a system of ideas intended to explain something, especially one based on general principles independent of the thing to be explained.

Threat versus Safety Theory (TVST) is the story of life, but more specifically humankind. Therefore, TVST is fundamentally about us—all of us.

I have included some information about myself and my background within the preface and postscript to add credibility and plausibility to the story. My inclusion within this story has left me somewhat conflicted. Although I know I wrote the story, it feels like I transcribed more than authored the story, and in no way am I central to this story.

On observation, life itself authored and performs this story for us every day. Humans simply need to attend the performance to understand. I hope this book gets our attention and gets us to attend to the lives we've been given with a greater intention to provide safety across the globe.

TVST is as basic as ancient wisdoms, yet as complicated as the universe itself. It is as compelling as anything that I have stumbled upon to date. The theory is as important to our health and happiness, and to global peace and tranquility, as any proposition or paradigm I have seen during my lifetime.

The premise of the theory is that chronic threat is the root of chronic disease, disability, pain, and suffering, and the provision of safety leads to health, vitality, connection, and happiness. However, it isn't so simple. Life itself begets threat. A condition of life is threat, threat is inevitable, and safety will never be a constant. Without the sensing of threat and the feeling of pain, the organic quickly becomes the inorganic—life dies. In this regard, pain is our friend and sentinel, not to be feared but to be respected and regarded. Pain signaling is essential to sustain life, therefore the goal of life should not be to eliminate the pain of life.

Recognize that fundamental to life is the protection and propagation of the genetic code. This is the purpose of life. To protect the code, we must be able to sense threat, then attend to it and act on it. Sensing threat is uncomfortable otherwise we would not be compelled to attend and act. Pain is a gift in this regard.

A key to life is to not get trapped in chronic threat, and thus chronic pain—this is suffering. Pain makes us aware of threat and then we need the support, skills, and strategies to escape to safety where we once again find our health, vitality, connections, and happiness. Thus, suffering becomes avoidable, and resilience is built.

This book creates a model of life through the lens of threat versus safety. This model, or paradigm, can be used in the formulation of an algorithm for better living and the thriving of the human species. Parts of the book challenge the status quo and the familiar. This may provoke unease, anxiety, irritability, and even anger—a threat response. Even the format of the book may feel unfamiliar and uncomfortable to the reader.

Fundamentally, the book is about the uncomfortable, the threats that come with being alive, the things we do to create maladaptive illusions of control in our lives, and the things that we can do to actually find safety and bring safety to the world.

Being uncomfortable and listening to our feelings—our eMOTIONS—gets us to move, to change, to do something different, and hopefully escape our threats and return to safety. The book is not meant to be warm and fuzzy, but to be real, yet hopeful, and ultimately helpful.

There are two cures for threat—safety and death. The second is a certainty of life. The first is preferred and accessible if together we try to achieve safety for ourselves and others. If chronic threat is the cause of chronic disease, disability, pain, and suffering, and safety and death are the cures for threat, then safety and death are also the cures for chronic disease, disability, pain, and suffering, yet safety is always just temporary. The skill to life is minimizing threat and maximizing safety. Whereas the art to life is learning to let go.

To successfully propagate the human genetic code, we must be safe. We must move to escape threat. Sometimes that includes the action of letting go. When caught in a rip tide, it is best to relax and ride the flow, as opposed to swimming against the current—and becoming exhausted. Don't be dissuaded by the unfamiliar and uncomfortable concepts, but absorb them, germinate them, flow with them, but don't fight them, nor cling to them.

TVST is not definitive, yet, and perhaps never will be—and you may be the one to write its next verse. TVST deserves deconstruction and reconstruction to make it better.

Enjoy the read,

DRC

Preface

"True love is being truly safe, seen,
and secure with another."

Ali F. Oeus

I ventured into medicine over forty years ago with a strong interest in exercise physiology, sports medicine, and human performance. Upon completion of medical school, I chose to pursue the specialty of Physical Medicine and Rehabilitation (PM&R). It is a specialty that has its roots in the treatment of chronic disease and disability, with the goals of not only treating disease, but treating disability to maximize function.

At the time I commenced my specialty training, I was far more interested in the treatment of acute injuries and the return to not only normal activities and play, but also the therapeutic and training techniques to maximize performance at an elite level.

PM&R had the foundational curriculum to set me up for this pursuit. In fact, the specialty has expanded in this direction over the years. It now includes the subspecialty of Sports Medicine, in which I am board certified, along with my primary specialty of PM&R.

However, along my path, I discovered that illness, disease, disability, pain, and suffering are far more interesting, perplexing, and significant in people's lives than the recovery from a strain, sprain, or broken bone. I came to understand that the absence of not just injury, but the absence of illness, disease, disability, pain, and suffering are essential for human performance and thriving. This is equally relevant to treating a world-class athlete and a homeless soul. Since that realization, I have never left the path of studying and treating illness, disease, disability, pain, and suffering in preference to injury.

I am still fascinated, and at the same time, disturbed, sometimes distressed, in this pursuit. I worry that in many areas, our models of disease and associated treatments are incorrect. We persist within these models, sometimes to the detriment of our patients. There are so many unanswered questions that need a better understanding and explanation. This is essential for a better model of healing, health, and wellness to emerge.

In medical training, we are made to feel as if we have arrived at the answers to these many questions—we haven't. Belief and dogma trump curiosity and questioning. Our patients believe we have the answers—we don't. We create a narrative that we do, and sometimes, we even believe it. Our hero story demands that we do know even when we don't know. Unfortunately, while we blindly persist on this failed path, the heartbreak of chronic disease, disability, pain, and suffering continues its expansion at pandemic levels. The many questions remain unanswered and frequently unasked.

Why do elite athletes get colds, migraines, or pull hamstrings around major competitions? Why are neck and back pain associated more with distress and job dissatisfaction than actual structural tissue injury? Why do chronic diseases of different tissues and organs tend to cluster together? What is the commonality for the cluster of obesity, diabetes, heart disease, arthritis, dementia, and cancer? Why are these diagnoses so prevalent in the undereducated, disenfranchised, and poor? Is obesity really excessive calories, fructose corn syrup, lack of discipline and willpower—or something more? Why would our body appear to attack themselves in a variety of "autoimmune" diseases? Why do our cells reject us to become cancer cells? The list goes on and on.

We as clinicians give, and patients accumulate, diagnoses as if these diagnoses are the long-sought answers to the questions. At times, these diagnoses serve more as anchors around our patients' necks when what they need is to be buoyed by understanding, hope, and a path forward to healing, health, and wellness.

It is not unusual for me to see patients with a vast assortment of both physical and mental health diagnoses. These frequently include *insomnia, anxiety, depression, addiction, headaches, allergies, asthma, dermatitis, arthritis, esophagitis, gastritis, colitis, constipation, endometriosis, obesity, and neck and back pain*. In some cases, the diagnostic list expands even further to include *bipolar affective disorder, post-traumatic stress disorder, borderline personality disorder, migraines, psoriasis, inflammatory bowel disease, inflammatory arthritis, cystitis, fibromyalgia, thyroiditis, diabetes mellitus, heart disease, seizures, and cancer.* The final stage in this expanding diagnostic process is the assignment of syndromes to the bigger mysteries in medicine—*metabolic, irritable bowel, chronic fatigue, adrenal fatigue, myalgic-encephalomyelitis, fibromyalgia, Gulf War, mast cell activation, multiple sensitivities, postural orthostatic tachycardia, polycystic ovary, Ehlers-Danlos, tethered cord, cervicomedullary, temporal-mandibular joint, post covid, premenstrual, toxic shock, acquired immunodeficiency, chronic (head, neck, back, abdominal, pelvic, myofascial...) pain—medically unexplained symptoms.* The list grows without answers to the questions as to the fundamental source of all these diagnoses. This list of possible diagnoses seems to be constantly expanding to no end.

Many of these patients have multiple clinicians from many specialties of medicine working with them. Each treats a diagnosis within their field of expertise. Or they chase ever-changing symptoms associated with a diagnosis from within their own silo, frequently without ever returning the patient to health and wellness. The diagnosis can be a false end point to which an errored algorithmic prescription is applied. It is a false answer, leading to a flawed treatment.

Additionally, clinicians are trained to look first for horses, common reasons for illnesses and diseases; if symptoms are unexplained then to look second for zebras, uncommon reasons for illnesses and diseases. Unfortunately, when still perplexed then some clinicians will look for and believe in unicorns. When settling into the imaginative and illusionary land of unicorns some real harm can be done to patients at a great cost. Every clinician wants to be the brilliant one who finds the zebra in amongst all the other horses. Most clinicians do not want to be the foolish one chasing or believing in unicorns. However, sometimes there is a very fine line between the pastures.

The etiology to most chronic disease, disability, pain, and suffering is not from the land of unicorns, nor zebras, but from the land of common horses—the land of common, chronic, cumulative, and compounded threats left untreated. When things don't add up before jumping into the land of zebras and unicorns clinicians should assess for threats. If there is no significant threat then the world of imagination and fantasy might be the best place for discovery of something new…to be confirmed prior to any application to patients. Belief is to be questioned, always.

A sobering statistic for clinicians is that clinical care is estimated to account for only 20% of health outcomes. The remaining 80% of health outcomes are felt to come from health behaviors and social determinants of health. Clinicians are not adept at addressing this other 80%. In many cases, we simply ignore the 80% to focus on the 20%. However, in other cases, clinicians simply lack the bandwidth and the power to affect the scope of care needs the patient has beyond the clinical dimension. Hope for our interventions and treatments being efficacious dwindles with this realization of inadequacy.

In addition, this 80% of the determinants of health may be too simply defined. In other words, all the components of the nonclinical factors that drive the health behavior and the social fabric of a patient may have yet to be understood. Are there more to the determinants of health than behavior and socioeconomic status?

Even within the 20% influenced by clinical care, for most chronic illnesses and diseases, our treatments aren't very good. Despite our pharmaceutical and technological advances, the incidence of chronic disease, disability, pain, and suffering is rapidly growing. Of great concern is the spread to the younger and younger populations. This manifests itself most notably in poor mental health, physical complaints, a variety of diseases, and early onset of cancers.

Given that clinical care is a small fraction of the determinants of health, and the treatments are frequently marginally effective, and even flawed, and infrequently cure, it seems reasonable to conclude we need a new paradigm in health care. We need a new assessment, both of what is right and what is wrong in clinical care. But perhaps more importantly, how do we successfully address 100% of the determinants of health? Better algorithms for treatment will follow a better paradigm for what it looks like to be a healthy and happy human being. We must consider:

- Is there a bigger picture behind every diagnosis?
- Is there correlation amongst diagnoses?
- Is there a common causation to explain the multitude of chronic diagnoses?
- If we dig deep enough, can we uncover the root of chronic disease and disability and the connection to pain and suffering?

PM&R has given me a platform to explore these many questions and more.

Our training in PM&R does include caring for high-performance injured athletes, young and old. However, more significantly, it encompasses caring for people who have had strokes, brain injuries, spinal cord injuries, amputations and the aged. It also comprises care for those who suffer from chronic neurologic disorders, such as multiple sclerosis, amyotrophic lateral sclerosis, movement disorders, dementias, and brain tumors, and those who suffer from chronic rheumatologic and fascio-musculo-skeletal disorders, including various forms of debilitating arthritis. The specialty covers a broad range of diseases and their associated disabilities. These, in turn, provide an expansive window and view to explore the many facets of human biology and associated physiology in what we characterize as illness and disease.

In addition to this broad view of humanity, our training required us to know everything about our patients—their chief complaint, history of present illness, past-medical and past-surgical histories; their generational, family, social, financial, and relationship histories; their educational, vocational, and avocational backgrounds; their home and work architecture, plus their personal habits. Then we obtained a review of all their systems. Finally, we were required to perform a detailed examination.

During my training, one of our attendings would famously ask the first-year residents on their first day, "is there a carpet in the home?" They never knew. He would scowl and say, "have that by tomorrow." The next day, they would report on the carpet in the home. Then he would ask, "how thick is the carpet?" They never knew. He would scowl and say, "have that by tomorrow." The next day, they would report on the thickness of the carpet, and he would ask, "what color is the carpet?" This was followed by a big grin. This would fluster the residents (and was a form of hazing). However, his point was you need to see people in their totality and within their environment to effectively treat them. Point made...and the carpet was green, for the record.

We cross-trained in internal medicine, neurology, orthopedics, rheumatology, pulmonology, cardiology, and even oncology, to be able to provide services across the spectrum of human maladies. These services are always offered with an eye towards our patients being able to maximally perform within their given environments. With this background, we are able to treat the toxins, allergens, antigens, pathogens, and injuries that afflict people. But most importantly, we are able to treat the complexity of a whole human being.

The specialty of PM&R uses a team approach to coordinating care with other clinicians, including doctors, nurses, psychologists, physical therapists, occupational therapists, speech pathologists, athletic trainers, vocational rehabilitationists, social workers, and case managers, to support the whole human being and maximize their function within their environment.

PM&R has been most widely known for its use of a team approach to health care and the focus on minimizing disability. But perhaps, what is most unique to our specialty is the inclusion of not just physical, but psychological, social, financial, and environmental perspectives in the care of our patients. It is a broad perspective that may not allow us to treat all the determinants of health, but it does allow us to appreciate the expanse of these determinants.

The concept of a true team within the specialty is invaluable for the delivery of care. Team dynamics are dependent on the cues and qualities of safety. For a team to work well, cooperation, adaptation, and compromise are essential. Social connection and bonding lead to easy communication. An environment of connection and safety increases cognitive functions, creativity, problem solving, judgement, and reliability. Recognition of each team member's skills, expertise, and contribution is facilitated within this environment. In addition, when things go wrong, this culture of safety builds transparency and resiliency.

A truly supportive team facilitates continuous quality improvement. Having a safe culture makes work and life more fun, but most importantly within medicine, alignment of purpose, mission, and vision within a safe culture leads to much better outcomes for patients. When care givers feel safe, seen, and secure, it is reciprocated and delivered within patient care. PM&R has been using these principles in the delivery of health care for generations. However, many within the world of medicine have failed to recognize the importance and influence of culture within the field when it comes to outcomes.

Unfortunately, the administrative and operational side of medicine has been infiltrated by virulent and contagious business and financial models and practices founded on conflict, competition, and maximizing returns on investment. These are threat-based models that ensure the survival of the fittest but fail to meet the needs of the many. These models can perhaps help with some efficiencies and resource allocation, but frequently fail when it comes to promoting the safety, health and wellbeing of patients, practitioners, and employees.

Hospitals and medical centers have become closer to bungee jumping companies than air traffic control when it comes to reliability and safety for both employees and patients.[1] Nearly a million patients in the U.S. are being permanently disabled or dying annually from misdiagnoses. Over 2.6 million deaths are caused by medical errors worldwide. In the U.S., 10% of all deaths are attributable to some king of medical error. It is estimated that around 500,000 medical errors occur annually in the U.S., making medical errors the third leading cause of death in the U.S. Over half of the hospitals in the U.S. are showing worsening patient safety trends. Maybe most worrisome is that over half of all practitioners are reporting burnout. Over 60% of all medical practices are now owned by nonclinicians. Infectious conflict, competition, and profit motivation are endemic within medicine in the U.S. and dominate the processes and culture of medicine. Meanwhile, people—patients and clinicians—are marginalized.

Lack of transparency and deceptive marketing and pricing strategies have contaminated both the hospital and insurance "markets" (notably markets to function effectively must be transparent and without deception). This costs patients their money and their health to the benefit and health of these corporations. The prioritization of the corporate species over the human species is incomprehensible, reprehensible, and just wrong. Humans have constructed and maintained the toxic and compromised institutions of their own suffering. Our blindness in and denial of our role in our afflictions are part of the problem of growing chronic disease, disability, pain, and suffering. Unknowingly, and sometimes, knowingly, we threaten our own species.

Private equity investment has over tripled in health care organizations within the last decade. Private equity is legally opaque and behaviorally irresponsible. Private equity's primary responsibility is to provide a return to investors, not to invest in the health of a patient. When profit and return on investment become priorities over wellness and health, there is a misalignment within medicine that creates over resourcing and utilization of high-profit procedures and under resourcing and utilization of low profit therapeutics. Clinicians, programs, and hospitals that cannot produce a desired return are shut down. Monies are not being invested in people and facilities to provide services. Return on investment is prioritized over capital improvements, salaries, and wages. Hospitals are left with excessive debt, while monies are shifted to the looters and profiteers.

With layers of corporate management and private equity firms pig piling onto an already suspect model of health care and finance, the infrastructure of medicine is collapsing under this heavy weight and messy malalignment. There are too many mouths to feed at the trough of medicine. Thus, wellness, health, outcomes, and safety are all being compromised.

Those at the bottom—the poor, disadvantaged, and underinsured—suffer more within this crumbling model of care.

Private equity owned hospitals use lesser trained lower cost labor, deny (ironically) equitable access to low-paying patients and low-paying diagnoses, have significantly worse rates of complications, worse quality measures, worse outcomes, and increased death rates compared with their peer non-PE hospitals. Cost cutting is a financial marker, whereas efficiency is a marker of quality. The two should not be conflated as being the same. Private equity firms don't create product, quality, better outcomes, nor money. They move money—to themselves and their investors.

> *"Given recent trends, we are concerned that some transactions may generate profits for those firms at the expense of patient's health, workers' safety, quality of care, and affordable healthcare for patients and taxpayers," wrote the Federal Trade Commission, the Justice Department, and the Health and Human Services Department in a joint request for information.*

Technically, the U.S. has a for-profit illness care model of medicine, marketed as health care. Healing and health are not just found by addressing the determinants and consequences of illness or injury, but by also addressing all the determinants of health. Absence of illness or injury is not the same as health. Health is predicated on being safe within one's environment. I've learned that the removal of threats allows us to survive, but the provision of safety allows us to thrive.

As a provider and patient in a care model that has seen the decline of health and longevity in the U.S. population, I am disturbed by our trajectory.[2] The United States of America now spends $4.2 trillion, ~ $13,000 per person, and 18% of our gross domestic product annually—roughly three times as much as our next peer country—chasing toxins, mutagens, allergens, antigens, pathogens, and structures within a broken for-profit illness care system. Despite our expensive high technology endeavors, we have only seen a decline in our health and longevity. We are number one in the world in spending on illness care, but we are over 50th globally in terms of overall health as a population. We are off target when it comes to healing, health, wellness, and happiness.

We probably rank even lower than 50th in the world in spending on actual health care, as opposed to illness care (this precise data is not actually reported). The lack of and disparity in spending on the real determinants of health is a major contributor to these grim statistics. TVST postulates that safety is the fundamental determinant of wellness and health, and we aren't investing to ensure we are safe within our society and the world.

Americans are living shorter and unhealthier lives because our system of illness care is mistargeted, as are our ideations, constructs, narratives, and beliefs of what it is to be a successful and meaningful human being. Health is predicated on being safe within the world. The provision of safety is the most meaningful thing we can do while we walk this earth. It facilitates our purpose here—to protect and propagate the human genetic code. Financial margins are important to survival, but maximizing profits to the detriment of safety and the provision of proper care is antithetical to the mission and to thriving.

Our better-quality health insurance is tightly tethered to employment, something hard to sustain when sick. Access to good insurance is difficult for the poor and the sick. Case in point: over 60% of bankruptcies involve the burden of medical bills.[3] As of 2021, over 30 million Americans remain uninsured in the United States. Contemplate, insurance is typically for random, infrequent, unpredictable, and sudden catastrophic events. Insurance applied to frequent, predictable, and slowly progressive processes such as chronic disease, disability, pain, and suffering seems inappropriate. There is a better way for covering American's health care needs.

In addition, the illness care system discourages people from accessing needed consultation and care through bureaucratic and financial threat. We must expand access to clinical care, improve efficiency in the delivery of care, act aggressively to control costs, and invest in societal support and services, along with financial and health equity, just to start.

Over 30% of American adults have multiple chronic conditions, such as two or more of the following: *allergies, asthma, chronic lung disease, diabetes, heart disease, high blood pressure, cancer, depression, anxiety, or another mental health condition.* Chronic diseases are the leading cause of disability worldwide and correlate with 80% of deaths in the United States. Chronic disease, disability, pain, and suffering are all much higher in the United States than its peer countries.

Over 70% of Americans are overweight or obese, and over 80% are metabolically unhealthy. Half have diabetes or prediabetes. Furthermore, the rates of allergic, inflammatory, and autoimmune diseases are rising at rates of 3-9% per year in the West. This is far faster than the speed of genetic change in the population. The U.S. has the worst rates among peer countries for life expectancy and death due to infant mortality, maternal mortality, obesity, assaults, and avoidable deaths.[4] The mortality gap between the U.S. and other high-income nations substantially expanded during the first two decades of the 21st century. Within the past five years, the gap has shown a dramatic upturn. [5] These are sobering statistics. Why are we seeing these trends?

Japan, despite their high rate of tobacco use, has the best rates of health and life expectancy, followed by the Scandinavian countries. It is notable that Japan is considered the safest country in the world. What might that say about the U.S.A?

Paralleling these health and longevity statistics, it also appears that brain volumes and intelligence within the U.S. are falling.[6] Are these trends correlated?

In addition, the U.S has fallen out of the top twenty happiest countries. These things are all connected.

"Our health care system squanders money because it is designed to react to [acute illness] and emergencies. Homeless shelters, hospital emergency rooms, jails, prisons—these are expensive and ineffective ways to intervene and there are people who clearly profit from this cycle of continued suffering."

Pete Early

I, in part, write the following based on my years of experience in the specialty of Physical Medicine and Rehabilitation and what it has taught me. The specialty has led me to see a new paradigm for disease, disability, pain and suffering, and their treatment. My traditional training and experience have also shown me the need to break down the silos within medicine. The explosion of information within medicine has only fertilized silo formation across medicine. Silos have even invaded my very specialty of medicine—a specialty with a historical strength and uniqueness in breaking through silos. These silos are getting narrower and narrower, and the walls thicker and thicker. They are effectively trapping us within. We seem unable to connect, turn around, or escape. We are positioned so close to the walls of our silos that we cannot see the greater picture—what is needed to be a healthy and happy human being. The allegory of Plato's Cave applies in that the symptoms we are chasing are only the shadows generated from a greater source and a greater reality. Chronic threat is the source and chronic disease, disability, pain, and suffering become our reality. We need to be able to pivot away from the shadows and towards the source, and then stand back to see the expanse of this source. It is then that we will be able fix this reality.

I also write this due to the distress I feel from the trends I see in the U.S. and the world, including, but not exclusive to, the declining health of the U.S. population. We have created a culture of conflict, competition, and threat. Whether it be in business, interpersonal relationships, broadcast media, social media, economics, politics, or foreign relations, we blindly engage in the propagation of conflict and competition, with not only binary outcomes—winners and losers, but unary outcomes—losers and losers.

Under the construct of exceptionalism, we repeatedly fortify a society and culture of threat without realizing the toxicity of chronic threat. Threat breeds not only more conflict and threat, but illness, disease, disability, pain, and suffering. This cycle expands and repeats itself. Threat is the escalating, unseen, and undeclared pandemic. Threat accumulates and compounds. Threat is expensive and wasteful. Threat overloads our illness care, criminal care, and social support systems. Threat disconnects us. Chronic threat is the biological root of our suffering.

Safety breaks this cycle, and not just on a philosophical or even moral level—but on a biological one. Safety also compounds. Providing equity and safety is not zero sum. No one is diminished for another's gain. Safety too is unary, but with winners and winners.

We and our institutions are failing in the building of a healthy society and culture. We are failing to recognize the importance of safety within both biology and culture when it comes to illness and disease versus wellness and health.

"Learn to lose well.
Account, Accept, Adjust, Adapt, and take Action
to be a winner.
That's an A-game.
The past is gone, the present is fleeting, the next play is now."

Lessons from Coach William McMahon

My medical career has been enormously satisfying to me, but no more so than my side job. I spent twenty-six years moonlighting as a football coach. The importance of the Xs and Os in football pale in comparison to the importance of culture. In my playing and coaching careers, I experienced cultures built on competition, intimidation, and punishment. But I also experienced cultures grounded in cooperation, collaboration, and accountability. The latter is a far more successful strategy and became foundational to the programs I was fortunate enough to lead.

Within this context, I was privileged to coach kids from all different backgrounds, orientations, religions, and heredities. My conclusion, paradoxically, is we are all the same. We all want to feel and be safe, seen, and secure in the world. Humans are very similar in physiology, appearance, emotions, behaviors, thoughts, and health when we are in threat versus safety.

Our threat responses are similar even to other species.

We also respond similarly when we are safe. Through time and evolution, the threat response (TR) and safety response (SR) have been programmed within our genetic code. Although our TRs and SRs are similar across the spectrum of threat through safety we are very different in our physiology, appearance, emotions, behaviors, thoughts, and health—we are changing all the time.

We tend to attribute bad behavior and poor academic performance to genetic, character, or brain deficiencies or defects. When in fact, cultures, societies, organizations, and environments of high threat activating a TR are the source of this perceived deficiency, disparity, and dysfunction.

It is notable that within our sameness, we are all constantly changing. The static-self, except as a construct of the mind, is biologically nonexistent for all of us. I saw kids who came to school with bad grades, learning disorders, and behavioral disorders change and demonstrate brilliance and leadership within an environment of safety. It gave me great pause to understand I was not special, but that my environment of relative privilege and safety accounted for who I was. All kids deserve the same. All humans deserve the same. Optimal human performance and health, whether on the gridiron, in the field of medicine, or in the field of life in general is dependent on a field, an environment, of safety.

"You are the universe experiencing itself."

Alan Watts

Humans and all life forms are ultimately connected to each other and to the environment. To practice medicine in a disconnected search for a toxin, mutagen, antigen, allergen, pathogen, or structural abnormality to cure chronic illness and disease absent the context of the environment is the ultimately folly. Most of our maladies are chronic illnesses and diseases; these chronic illnesses and diseases do not arise from a single entity. Rather, they are derived from a systemic coordinated cascade of phenotypic and physiologic changes. It is our environment interfacing with our genetic code that determines these phenotypic and physiologic changes, and subsequently, our state of illness and disease versus wellness and health.

Evolutionary biology dating back more than 4 billion years shows us that all organisms respond and operate differently, depending on whether they are in threat or safety within their environments. We are no different in this regard than the most ancient and simplest organisms. We change our phenotype and physiology when in threat versus safety. Every cell changes in threat versus safety. Our entire system shifts in threat versus safety. This reflects a coordinated effort to try to survive, hopefully thrive, and ultimately, to protect and propagate the species.

In the following, I have used my breadth of experiences and evolving perspectives to explain some of the multitudes of change that occur within threat versus safety. In addition, I have attempted to deconstruct silos to look around and see the commonalities of chronic diseases. Finally, I have even attempted, through the wide lens of TVST, to explain the human condition itself.

Threat is the source of most chronic disease, disability, pain, and suffering (is Safety Deficit Disorder (SDD) the proper diagnosis?).

The following is a theory based on my experiences and the available science. It is driven by curiosity, relies on an element of creativity, and is anything but definitive. I hope my musings can serve as a starting point to a new and different discussion regarding illness and disease versus wellness and health, and their determinants, without being too threatening in the questioning of the status quo. But above all else, I hope this helps to build a logical and efficient treatment algorithm for chronic disease, disability, pain, and suffering. And in turn, I hope this will help restore individual human beings and the world to wellness, health, and happiness.

"What is spoken of as a clinical picture is not just a photograph of a sick man in bed; it is an impressionistic painting of the patient surrounded by his home, his work, his relations, his friends, his joys, sorrows, hopes and fears. Now all of this background of sickness which bears so strongly on the symptomatology is liable to be lost sight of in the hospital."

Francis Peabody

Guide to TVST

TVST

"Humankind has not woven the web of life.
We are but one thread within it.
Whatever we do to the web, we do to ourselves.
All things are bound together.
All things connect."

Chief Seattle

Threat versus Safety Theory states that the root of illness and disease is threat, and the root of wellness and health is safety. The theory is not specific to physical illness and disease versus wellness and health. It is inclusive of emotional, behavioral, social, mental, cultural, and spiritual illness and disease versus wellness and health. Of major concern is that most of humans' afflictions are not from acute threats, but from chronic threats. More specifically, the hypothesis is that being trapped in chronic threat is the root of chronic disease, disability, pain, and suffering.

The foundation of the theory is grounded in evolutionary biology and physiology. Biologists usually define life as an entity that responds to its environment, has a boundary, metabolizes substrates, consumes energy, grows, regenerates, and reproduces itself. Perhaps this definition should include defends itself, too.

Under this definition, life first formed on Earth over 4 billion years ago. Life initially was comprised of simple single-celled organisms, archaea and bacteria. More complex life forms did not evolve until over 2 billion years later, and the human species, specifically Homo sapiens, are just 300,000-250,000-years old. Traditionally, humans are thought of as hairless, upright, bipedal great apes, with highly evolved social, intellectual, language, and conscious functions that for most of human history lived as hunter-forager-gatherers. TVST also uses a perspective that humans are the multi-cellular, differentiated-cellular, descendants from ancient archaea cells, with the adaptation and cooperation of endosymbiotic and exosymbiotic bacteria.[7]

To fully appreciate and understand the human species, knowledge of ancient simple single-cell life forms is foundational. Understanding their responses to threat and safety is enlightening. Of note, there is some conservation of these traits along the diverging evolutionary paths of all organisms. Additionally, when looking at cellular biology and physiology, there is a deep connection to the environment that influences all cellular functions. The fundamental environmental context that cells respond to is whether they are in a state of threat or a state of safety.

Life

About 10 (perhaps 20+ based on more recent estimates of the universe's age) billion years after the creation of the universe cellular life arose within the earth's primordial inorganic soup. How this transpired is still unknown. The specific ends also remain unclear, but what is clear is that the goal of life is to protect, defend, sustain, and ultimately propagate its genetic code.

A primary requirement for cellular life is a boundary, a selective semipermeable membrane, to differentiate the life form from its surroundings—to differentiate the organic from the inorganic.

There is strong evidence that fatty acids can spontaneously form phospholipid bilayers within soups. With the infusion of gases into the soup, fatty bubbles form that can easily and repetitively capture substrates (sugars, amino acids, nucleic acids) in what 4 billion years ago could have been inorganic cell-like prebiotic containers (protocells) that could serve as precursors to organic life forms.[8,9]

Amino acids can spontaneously form chains to create small peptide molecules that can congregate and self-replicate, and nucleic acids can spontaneously form chains that that also congregate and self-replicate—all within the inorganic world. These functions mimic the essential functions of cellular life. [10] Perhaps, this suggests a form of protointelligence in matter or materials not considered yet to be alive.

All the nucleic acids and sugars that make up the genetic code (RNA, DNA) of all life forms have been found within comets, asteroids, and meteoroids from outer space. In fact, there is evidence that amino acids from outer space were forming proteins and folded formations 4.6 billion years ago, 400 million years prior to the first life forms on Earth.

Cosmic dust can spontaneously form carbon and hydrogen cage like structures that may have had a role in protecting prebiotic molecules from the toxicities of the universe. Given all these inorganic functions in some sense, biological evolution seems to have predated life itself—or at least how life has been defined.[11]

The next requirement of cellular life after a boundary is the ability to manage energy—a metabolism. This is vital for performing the various basic biologic functions of life. These include movement, defense, growth, repair, regeneration, and reproduction.

The primitive world was full of extreme energies. These included radiation, electromagnetism, vibration, heat, pressure, explosions, vents, currents, and acid and alkaline flows. These energies only needed to be coopted, detoxified, harnessed, and organized within cellular life forms to run the metabolic functions of cellular life.[12] Exactly how this happened is unclear, but human physiology is an organized reflection of these primitive energies.

The universe is full of free energies, yet life's survival is dependent on processes to manage and minimize these free and potentially toxic energies. Life needs a program for these functions. Life needs a program for defense, metabolism, growth, repair, regeneration, and reproduction. Life needs a code.

The genetic code for life on earth is made from chains of nucleic acids which were readily available within the primordial soup. These molecules may have spontaneously formed on Earth within deep hot ocean vents or around volcanic eruptions. However, more likely, the major source of these building blocks for life was from outer space on comets, asteroids, and meteoroids that arrived during the great bombardment period of Earth's history, along with water in the form of ice—another important ingredient for the soup of life.[13] Notably, the other building blocks necessary for life including amino acids, lipids, and sugars can all be formed and found within the environment of outer space, too.

(One must also consider that much of life isn't directly driven by the nucleic acid code. RNA and DNA primarily code for peptides and proteins and it is the spontaneous nature (or perhaps intelligence) of things like electricity, ions, salts, water, carbohydrates, and fats that account for much structure, function, appearance, and phenotype of lifeforms. Nevertheless, the genetic code has a dominant role in governing life and life forms.)

Beyond these basic requirements of a boundary, metabolism, and code cellular life forms had the essential need for a defensive system to protect themselves from the many threats within the environment. Indeed, many of the primordial threats to cellular life were the same as the many energy sources of the planet. The primitive archaea and bacteria detoxified, coopted, and managed some of these energies for the functions of life. The integration of fuel acquisition and energy production for cellular use (metabolism) and defenses (immunities) became, in part, both the defense against the threats of these extreme energies, as well as the networks to convert and produce energy to run these primitive cells.

In addition to the inherent toxicities of the planet, life itself begot threats in the form of other life—competitors and predators.

For survival, cellular life required successful metabolic and immunologic responses to threats. A *threat response* (TR) starts with the sensing—cellular awareness—and then the signaling—cellular communication—that threat is present. This threat sensing and signaling (TS) dictates threat-related changes in the expression of the genetic code. Reciprocally, a *safety response* (SR) starts with the sensing and signaling (SS) that the organism is safe within its environment that results in different expressions of the genetic code.

These responses involve adjustments in genomic formation, epigenetic modifications, and subsequent code transcription and translation. The adjustments lead to phenotypic transformations in cellular production, function, and appearance. [14] In humans, this transformation not only includes these cellular changes, but also changes in systemic physiology. These changes further affect how humans, as whole sentient, social, cognitive, and conscious beings, function, appear, feel, behave, and think.

When in a state of threat, all life forms, including mitochondria (descendants from and remnants of endosymbiotic bacteria that became the defense and power plants within human cells[15]), cells, tissues, organs, and whole beings change. Mitochondria, cells, tissues, organs, and whole beings are different when in a state of threat versus a state of safety. The genomic structure, epigenomic pattern, and coded transcription and translation adjust to the changes sensed within the environment, with subsequent corresponding phenotypic and physiologic transformation along the spectrum from threat to safety.

TVST purposes that it is the interface of the exposome—variable state of the environment—with the genome—variable expression of the genetic code—that by and large determines how organisms present themselves within the world. This, of course, includes humans. Contemplate the fact that the blueprints for the construction of one human being requires only a meter of DNA—containing just over 20,000 genes within one tiny cell. The builder is an infinite universe.

The Center for Disease Control (CDC) defines the exposome as a measure of all exposures of an individual. TVST expands this to include all exposures of the cells of an organism. Thus, both the external and internal environment are at play (exposome = the exterosome + the interosome). This has implications in the areas of emotion, behavior, and thought regulation and management that influence the components and mixture of the physiologic soup bathing human cells. Self-generated internal inputs should not be lost within these calculations of the environment and exposures.

Part of TVST postulates that as there are nuanced differences in genomes, individuals within a species vary a bit. Therefore, there are nuanced responses to specific forms of threat versus safety. However, overall, the TR is very similar within organisms, cells, tissues, organs, and species. All species change their metabolism, and create toxins, oxidants, and inflammation within a TR. More uniquely, large differentiated-cellular species have neurologic and vascular structures to communicate threat broadly across the cells within the greater system. The autonomic nervous network and the hypothalamic-pituitary-adrenal neurologic-endocrine-vascular network functions are remarkably conserved across more modern species.

As well as individual gene expression, there are gene clusters within the genome that are coordinated in expression to affect a desired integrated response to environmental changes. There are many common nonspecific changes to the expressed nucleosome, transcriptome, proteome, metabolome, inflammasome, connectome, biome…the phenome regardless of the specific type or level of threat versus safety experienced. These changes match the needs of the moment from threat through safety. The threat-related genomic clusters coordinate the components of a systemic TR. These associated phenomena help explain why stimulation of the physical, emotional, social, mental, or spiritual networks can result in such similar systemic physiologic expressions and responses when in threat versus safety.

Yet, humans don't all react in exactly the same manner, with the same patterns of illness and disease, when under threat. There are clearly some genetic variations and predispositions to certain illnesses and diseases for individuals and their relatives. TVST does acknowledge that genetic variability expressed leads to a multitude of symptoms and diagnoses that present differently throughout the human population. But more importantly, the environment is vastly variable. As the interface of the exposome with the genome creates the phenome, this may be a better explanation of the variable outcomes and the multitude of threat-related diagnoses than simply genetic differences.

TVST also contends that the root cause of chronic disease, disability, pain, and suffering is the same. Although the presentations and diagnoses may differ, the source does not—chronic threat.

In addition, it is notable that the human TR is biased towards the responses needed in a physical attack, such as a pathogen or predator, even when experiencing an emotional, social, mental, or spiritual attack. This makes some sense. Physical threats were the primary concern of single-cellular life forms for billions of years within the primitive world long before the evolution of more complex emotional, social, mental, cultural, and spiritual networks, and humans. In addition, physical threats remained the primary concern of Homo sapiens for the vast majority of their 300,000-250,000-year history.

Threat Phenotypes

In a response to threat, organisms selectively express their genetic code with the primary goal to physically survive—resources are prioritized for a defense.[16]

Ancient archaeal and bacterial species have many responses to threats. They can swarm to attack or colonize within biofilms to protect as methods of defense.[17] Archaeal and bacterial colonies are dynamic. They possess quick and efficient intercellular communications, specialization of functions, sharing of resources and information—including code, and even sacrificing of structures—to the point of programmed death. Colonies offer the power of numbers, organization, and diversity, as well as the power of collective awareness and intelligence. These properties facilitate organized deployment of defenses to protect and propagate the genetic code forward into the next generation.[18] Intra-species inter-organism communication was an early and essential tool for the warning of threat and the coordination of a group defense strategy.[19] It is important to note that sociality is very primitive, a trait as old as life itself.

Additionally, archaea and bacteria can fight by various means. They can secrete toxins, penetrate, or engulf, and enzymatically destroy a competitor or predator.[20] Motility or mobility is also a great asset in a defense. The simple ability to flee from a toxin, pathogen, or predator is a very efficient defense strategy. Archaea and bacteria can also lower their metabolism and hide, even becoming dormant (for more than a hundred million years[21]) when threat is severe. Rapid reproduction by fission (simply dividing in half) is a method to not just propagate, but to protect the genetic code within numbers. (Once safe, sociality returns and can include fusion which recombines multiple organisms into one, too.)

Included in alternative primitive defense programs was not only dormancy, but also programmed cell death or apoptosis—sometimes referred to as altruistic cell death to connotate the sacrificial and social nature of this function in times of threat. Within these primitive cellular life forms, there is a coupling between immunity (defense), metabolism (energy), and programmed cell death (sacrifice). As a result, a cell chooses one of these defense programs by sensing the degree of threat and damage present. These programs incorporate both options of individual survival and species survival to ensure the defense, protection, and propagation of the genetic code.[22]

Archaea and bacteria engaged in incessant strategies for survival and evolved to have mechanisms that are now reflected in the modern innate, adaptive, and surveillance immune and metabolic responses of human cells.[23]

(Note: Human mitochondria, intercellular endosymbiotic bacterial descendants and remnants, undergo fission and become more aggressive, inflammatory, oxidative, and catabolic in threat. They then undergo fusion and become more restorative, anti-inflammatory, and anabolic once in safety. All this occurs within the cytoplasm of human cells. It is akin to the way primitive bacteria perform these actions in threat versus safety. Multicellular organisms lack this flexibility, but it is conserved within their mitochondria. Mitochondria also show social behavior much like their ancient bacterial ancestors. This includes fusion, communication, sharing of information (code), group formation, synchronization, interdependence, specialization of function, division of labor, and sacrifice for the greater good. Wildly, under conditions of threat mitochondria will export their DNA into the nucleus of the cell to be spliced into the genome to change the genetic code—which has all kinds of implications for illness and disease, especially cancer. Mitochondria serve as sensory organelles for the cell, perhaps akin to the brain serving as a sensory organ for the body—sentinels for the whole.)

Humans have very similar defense programs to their ancient ancestors. This makes sense, given that humans evolved from these primitive and simple life forms. Humans form tribes and attack via mobs, gangs, militias, and militaries, as well as protect each other within colonies. Human colonies similarly offer the power of numbers, organization, and diversity, as well as the power of collective awareness and intelligence. (Arguably, a human is just such a colony of cells in this regard.) These social structures and functions require communication, specialization, sharing of resources, sharing of information, sacrifice, and even death, all within the deployment of an organized defense.

From an evolutionary standpoint, the goal should be to survive and thrive, but most importantly and primary is to defend, protect, and propagate the genetic code into the next generation. However, humans can sometimes miss the mark regarding the fundamental principles and functions of a species. As a result, at times, humans threaten their own survival as a species. The greatest threat to the human species *is* the human species. This is an ironic biological anomaly, perhaps analogous to a cancer cell within the body.

Humans can attack or run when under threat. Humans can isolate and hide in severe threat. In extreme threat, humans, too, can collapse and appear dormant when feigning death, fainting, or slipping into shock. Repopulating is a way of protecting and propagating the genetic code within numbers for humans, as well. However, humans are relatively infertile when in threat states. Rapid reproduction is not an option.

Whole human beings in a defense state are doing all these things while their individual mitochondria and cells are participating in these defense processes, as well. This is undertaken in a manner very similar to the defenses of their primitive bacterial and archaeal ancestors in the primordial soup. Human mitochondria and cells, most notably white blood cells—but all cells to some degree—participate in innate, adaptive, and surveillance immune responses. Human cells can also choose relative hypometabolic dormancy (examples include senescence, "immune-quiescence," "immune-senescence," "immune-paralysis," or "immune-exhaustion.") or programmed cell death within a defense response.[24,25]

"I will pick up the hook
You will see something new
Two things and I call them Thing One and Thing Two
These Things will not bite you
They want to have fun
Then out of the box came Thing Two and Thing One"

The Cat in the Hat – Dr Suess

There are multiple reactions to and programs for threat. Each of these threat states involves different interpretations of the genetic code, expressed in different phenotypes and physiologies. Grossly, the TR is biphasic in appearance, as characterized by the Threat-1 (T1) and Threat-2 (T2) phenotypes outlined below.

TVST characterizes these threat states as:

- *Threat-1a: Fortress – colonization*
- *Threat-1b: Fight*
- *Threat-1c: Flight*
- *Threat-2a: Freeze*
- *Threat-2b: Falter*
- *Threat-2c: Faint*

Threat-1 (T1) phenotypes are notably relatively active states, with high energy production to facilitate mobilization and other defense responses. In T1 feeling "out of my mind" is common.

Threat-1a is a more socially protective state than other threat states, yet should be recognized as a defensive state with associated physiology.

Threat-2 (T2) phenotypes are notably relatively inactive states, with progressive energy conservation and immobilization to slow degeneration and allow for preservation. In T2, feeling "stuck in my mind" is common.

Threat-2a is a transitory state of immobility when threat assessment occurs while maintaining a relatively high metabolic rate. A survival strategy is then deployed from the multiple options outlined.

Threat can be resolved by either defeat or retreat. Unfortunately, injury and death can be the sequelae of an attack. Additionally, mitochondrial, cellular, tissue, and organ degeneration and destruction can be the other sequelae of a TR.

Safety Phenotypes

Once in safety, life forms express their genetic code to thrive—resources are prioritized for construction and reproduction.

When a threat has been resolved, life can devote resources to acquisition, accumulation, recovery, repair, regeneration, growth, and abundant reproduction. Biologically, resilience is predicated on periods of restored safety.

Safety has multiple states as well. Each state involves different interpretations of the genetic code, expressed in different phenotypes and physiologies.

The colony functions to facilitate these states. In safety, reproduction is a priority. So too are the acquisition and accumulation of resources. Also, periods of decreased activity and decreased resource utilization are essential for growth, repair, regeneration, recovery, reproduction, and thriving. Similarly, the SR is biphasic in appearance, as characterized by the Safety-1 (S1) and Safety-2 (S2) phenotypes outlined below.

TVST characterizes these states as:

- *Safety-1a: Bond – colonization*
- *Safety-1b: Breed*
- *Safety-1c: Feed*
- *Safety-2a: Digest*
- *Safety-2b: Nest*
- *Safety-2c: Rest*

Safety-1 (S1) phenotypes are notably relatively active higher energy-consumption states that include socializing, playing, acquiring, creating, constructing, laboring, foraging, feeding, and fornicating.

Safety-2 (S2) phenotypes are relatively inactive states, with progressive energy conservation and solitude to allow for recovery, repair, regeneration, and growth. These are intrinsic states to most life forms when out of threat and when other needs have been met.

People want to be resilient. Examining resilient people demonstrates the qualities of connection, vulnerability, perseverance, gratiosity (gratitude), and possibility.[26] The reflexic aspirations and goals are to try and achieve these qualities. However, the reality is that when these things are created ideations, cognitive constructs, symbolic narratives, and held beliefs of the mind, and work to be performed, goals to be achieved, aspirations to be conquered, or desires to be quenched, then this striving can add to the pain of inadequacy and increase threat. These qualities are phenotypic to safety and naturally come to life within a world of safety. Time in safety is resilience.

In humans, cultural constructs can interfere with the S2 states. S2 states are necessary and allow for thriving. Being in S2 requires becoming comfortable within the world and letting go of the need to do. Relaxation, meditation, prayer, and sleep should be sacred and welcomed as biologically necessary activities and not demeaned or marginalized.

Moving from human doing to human being is to follow the laws of nature. This process is necessary for the restorative functions of growth, repair, regeneration, and reproduction. The necessity of the S2 states demands rethinking the modern human's version of what it means to be a successful, meaningful, and purposeful person versus a healthy organism within present culture. Chronic work, production, and competition, in a search for profit, prestige, power, and achievement are overvalued in Western and medical culture. By contrast, contemplation, creation, recreation, relaxation, restoration, and sleep are undervalued. We equate relaxation and sleep with laziness and weakness. This is a setup for lost resilience and fertile grounds for disease, disability, pain, and suffering.

There is no better example of this than within the culture of medical training. There is little time allowed for relaxation, restoration, or sleep. Students and residents who take this time are marginalized, if not hazed. These students and residents are frequently "weeded" out of programs. Conflict, competition, and the survival of the fittest are at play within this paradigm. It is felt this process will produce the best of the best. However, this paradigm only leads to arrogance, aggression, disconnection, irritability, inflexibility, anxiety, obsessiveness, depression, despair, burnout, and even suicide. This problem has improved over the years, but fundamentally, this construct is still endemic in the culture of medicine.[27]

It should not go without mentioning that intellectual creativity and capacity (IQ), as well as emotional empathy, compassion, and connectivity (EQ)—both neocortical prefrontal lobe functions—fall throughout the medical training process (and are falling across the general population of the United States).[28] Systems get exactly what they are designed for—the perfectly constructed physician for conflict, competition, and burnout, devoid of the ability to actually care or care for. This paradigm, grounded in the idea of working until you drop, is also endemic within the U.S. and most of western culture.

Allowing and valuing S2 phenotypes may require building new ideations, constructs, narratives, beliefs, and paradigms of health and wellness within all cultures—especially within medicine and in those who are supposed to guide us to health and wellness. The S2 phenotype demands devoted time and safety.

Humans must be physically, emotionally, behaviorally, socially, mentally, culturally, financially, and spiritually safe to be healthy and well. Humans all need a safety net that they cannot fall below, that protects them from a hard landing in times of threat. Humans need clean water, pure air, healthy food, safe housing, secure finances, expressive freedom, close social connections, positive cognitive constructs, reintegration with the natural world, access to healthcare, and finally—time to rest and recover.

"Our body and mind have the capacity
to heal themselves if we allow them to rest.
Stopping, calming, and resting are preconditions for healing.
If we cannot stop, the course of our destruction
will just continue."

Thich Nhat Hanh

To offer some final comments on these threat and safety phenotypes, it should be noted that they represent major observable shifts in physiology. However, physiology is a flow with constant change. Physiology is never a fixed state. Cellular life is not a constant, and even these classic phenotypes are not fixed physiologies or specific points. The exposome—the interosome and exterosome—determine the phenome—the part of the genome expressed—on a sliding scale from threat to safety, within a state of constant change. Equilibrium is, therefore, only a relative concept and never an absolute.

The Illusion of Stasis

The concept of *homeostasis* (a self-regulating process through which biological systems maintain stability) also allows for subtle changes in physiology but suggests a fixed equilibrium to which the organism has a goal to achieve—a set point.[29] An analogy to consider is a seesaw with a potential balance point. This balance point may be hard to achieve, yet the action is centered around this point. If the system shifts one way, then there are feed-forward and feed-back loops to pull the system towards the balance point. If the system shifts in the other direction, then different feed-forward and feed-back loops will pull the system towards the balance point.

These types of loops do exist within organic systems, but this analogy is inadequate to explain all human physiology. This dated construct of homeostasis is an illusion that suggests the organism has an idealized center point—normal—and co-ordinates itself around this idealized center point. This center point is nonexistent, and thus there is no normal. The organism has infinite phenotypes and infinite physiologies that are in a state of constant change in relation to the interosome and exterosome and the state of threat versus safety.

Granted, cellular life has physiologic parameters required for survival, and cellular life does the best it can to stay within those parameters. But there is no set equilibrium or balance point within those parameters. There is just the physiologic spectrum of life—just change. Homeostasis as a century-old concept needs re-evaluation.

"No man ever steps in the same river twice,
for it's not the same river and he's not the same man."

Heraclitus

The term *allostasis* implies a deviation from the homeostatic point, a deviation from equilibrium, or perhaps the reset to a new equilibrium. It may be an unnecessary term as homeostasis, in fact, allows for an "other point."[30] However, neither the concept of homeostasis nor allostasis encompasses the fluid continuum of genomic expression along the spectrum of threat through safety physiology that exists without a point of true equilibrium. Perhaps terms such as *menacufluent*—threat flow—and *salvufluent*—safety flow—would be more applicable to describe the spectrum of physiology organisms demonstrate over time.

Allostatic load is defined as "the wear and tear on the body," which accumulates as an individual is exposed to repeated or chronic distress and away from homeostasis.[31] This concept has some validity within TVST. However, TVST is more concerned with the threat load (TL) that results in the spectrum of phenotypic and associated physiologic changes thus seen as the wear and tear on the body. The wear and tear are the result of chronic threat. This is manifested as disease, disability, pain, and suffering. Therefore, the wear and tear on the body are the afflictions, not the causation of the afflictions. Also, the wear and tear extend beyond the body to include the tattering of the emotional, behavioral, social, mental, cultural, and spiritual fabric of an individual and a society.

The concepts of homeostasis and allostasis work for some things, but not the whole (such as classic versus quantum physics). New paradigms that have moved beyond the concepts of homeostasis and allostasis will be of value for the fields of physiology and medicine in the 21st century.

In addition, consider this: the system has multiple environmental interfaces and inputs that complement, compound, or counter each other. A wound on a leg can activate a TR, while a kind, compassionate, and warm friend can activate a SR to counter the TR. Indeed, wounds heal faster when people feel safe.[32] Compounded threat does the opposite.

A physical wound, compounded with an emotional or social wound, can delay healing. People with more threats and distress in their lives have more disease, disability, pain, and suffering. They take longer to heal. In addition, they have worse outcomes after surgery. Socioeconomic status is the best determinant for outcomes after surgery. Poverty is a predictor of complications, poor healing, infections, illness, and disease.[33]

The multiple interfaces with the human internal and external environments resulting in genetic and epigenetic changes add much complexity beyond homeostatic loops that must be considered in human physiology. To be stuck in threat is to be stuck in threat physiology, with no internal pull towards any idealized and illusionary homeostatic point.

Threat and distress are unavoidable. To be fully alive is to experience states of threat, and hopefully states of safety. However, to have chronic disease, disability, pain, and suffering—physical, emotional, behavioral, social, mental, cultural, or spiritual—is to be stuck or trapped in the threat states. Chronic threat is the root of chronic disease, disability, pain, and suffering.

"Truth is rhythmical: if it implies stasis,
it is platitude."

Robert Grudin

Philosophy

From a philosophical perspective, the ever-changing physiology, with a lack of stability or normalcy, raises the question as to whether a static "true" or "authentic" self can possibly exist within this constant change—the only constant being change. The real-self is one of constant change.

The colloquial true self is a construct, an idealized illusion. Humans are different depending on threat versus safety—bad-good, antisocial-social, reactive-contemplative, catabolic-anabolic, etc., or somewhere in between. TVST contends that no human is pure bad unless they are always threatened, and no human is pure good unless they are always safe. Evil—a devil, or even an angel—is all pure construct. Heaven and Hell, Good and Evil, are relative to context and contextual to safety versus threat on Earth.

"The time to make up your mind
about people is never."

Katharine Hepburn

The mind and the body shift, one and the same, together within the physiologic soup. It is a coordinated systemic response to the interosome and exterosome. The mind in its separateness is a false ideation, construct, narrative, and belief. The dissolution of a mind-body duality is clear at the level of the ever-changing soup. Consistent with infinite phenotypes and physiologies is the realization that humans have ever changing and infinite bodies, infinite minds, and infinite selves.

The constructs of homeostasis and of an authentic static-self contain an element of human hubris. They appear to be egocentric illusions of an idealized whole that is in some way in control, in charge, and goal-directed towards a point of metabolic, physical, emotional, behavioral, social, mental, and spiritual stability and normalcy. Biology does not support this construct.

Humans also tend to visualize the brain and the mind as being one and the same. Humans infer that the master control center for the system is the brain. Humans suspect it is within the head, somewhere behind the eyes and between the ears, where the mind and the self reside. From this cockpit, a good pilot can achieve and provide systemic and environmental control.

Human awareness, cognition, memory, and intellect lie within the cortex of the brain, perhaps more specifically within the neocortex (phylogenetically, the brain's newest part) of the brain. Yet, these functions are never separate from the system and cellular awareness, memory, and intellect, which are diffuse. And most significantly they are never separate from the interface of the ever-changing environment and the influences of the exposome over the genome. The mind and the self are not only in constant change but are also diffuse.

Cellular awareness and intellect is readily apparent within laboratory created human cell organoids and biobots that although disconnected from the whole of a human being demonstrate phenotypic changes, problem solving ability, and purposeful activity.

Planaria, slime molds, sea anemones, and octopuses diverged from humans along the evolutionary tree over a half a billion years ago, and from each other within 100 million years ago. All these species demonstrate that a master brain is not a requirement for awareness, memory, intellect, or control.[34,35,36]

Planaria can not only survive after losing their heads but can regenerate with memories intact. Perhaps even more disconcerting is planaria demonstrate phenotypic changes from environmental stimuli such as electricity or chemical baths that are not coordinated through the genome and epigenetic signaling. No central control from the brain nor the genome—wild.

Slime mold cells cooperate, divide, and conquer when in threat to defend against attack. These same cells cooperate, coordinate, and collaborate when in safety to divide tasks such as detecting and acquiring food, sharing of resources, and moving into a colony. In addition, all multi-cellular species' have immune cells that remember and attack pathogens they have seen before.

Octopi have a nervous system that is disseminated within their many appendages and cognition occurs locally within each appendage to feed back to the greater system. Are Homo sapiens that different? Humans do have a massive cognitive organ within their heads, but the gut has its own enteric nervous system, and the heart does as well. How much of human awareness, processing, and intellect is driven from these other organs—do we really know?

Nonneuronal cells can and do store information. Approximately, 3.7 billion years of evolution preceded neurons and brains. Neuro-cognition came late (0.5 billion years ago) and human neocortical cognition came last (just 0.00025 billion years ago). Life outside of the human neocortex tends to come one moment at a time, the present moment, but not completely as memories of the past are preserved throughout the system within the cells beyond the brain.

Cells have purpose and goals to achieve—to protect and propagate the code and life. They signal to each other through electromagnetism, vibration, molecules, and chemicals. Brain organoids even exhibit these properties as these cells self-organize into systems, too, without any greater oversite from a body. The act of living, life itself, by default is a conscious and intelligent state. All consciousness and intelligence are a collective and connected. The concepts of the mind, and even consciousness, seem to align better with physiology—perhaps even with the concepts of energy within physics—than with anatomy.

The neural tree's evolution commenced approximately a half a billion years ago, preceded by 3.7 billion years of cellular evolution. The Homo sapien brain is under 300,000 years old. The brain is just a part of the neurologic, endocrinologic, and immunologic networks that are all part of a greater system. The brain and neurologic network has an important role in quickly coordinating the other cells and the greater system over large distances. Nevertheless, the brain is not a master controller, but a coordinator, and neither is the mind a master controller. The environment is the master controller to which the system—the mind-body inclusive of the brain—adjusts.

Under the lens of TVST, these concepts of stasis, stability, mind, static self, and normalcy dissolve away. The whole, the sum of the parts—all the cells, and all the cell's parts—in a dynamic relation to each other ultimately define the ever-changing organism called a human.

Under threat, the system can undergo dissolution from the cortex of the brain—from cortical awareness, memory, and intellect—all the way to the level of the cells and parts of the cell—to cellular awareness, memory, and intellect. In addition, all these parts and their sum are never separate from the environment. Thus, they always reflect the ever-changing environment, resulting in infinite physiologies and infinite loci of control.

It seems highly probable that the basic coding for the sense of self (as well as the sense of the other and even things like hate and jealousy or empathy and altruism) began in primordial single-celled organisms 4 billion years ago. Single-cellular life included colonized life that would eventually evolve into multi-cellular life. Multi-cellular life eventually led to differentiated-cellular life, with tissues such as muscles and bones, and organs such as pancreases and brains, finally evolving to the apex species of human beings. All these stages represented major evolutionary leaps. Yet these leaps conserved the functions of the cells that came before them—including their sense of self.

The self within this context became progressively larger and more complicated with each additional cell within an organism. These metamorphoses resulted in the self being more than something to be protected and defended. At the apex, the self became something that could be contemplated and expressed as a created ideation, cognitive construct, symbolic narrative, and held belief, to then be idealized and idolized by the most modern species of human beings.

"Our terminal decline into old age and death stems from the fine print of a contract we signed with our mitochondria two billion years ago."

Nick Lane

A brief reflection: Mitochondria, descendants and remnants of endosymbiotic bacteria that power and defend the cell, have a sense of self and protect and propagate themselves not in a multi-cellular fashion, but in an endo-cellular fashion. Each of a human's trillions of cells has thousands of mitochondria. In other words, thousands x trillions of selves within selves within a self. And they're all ever-changing in phenotype and physiology, depending on their relatively simple genome interfacing with the complexity of the exposome…the complexity of the universe. How distressing would it be to be a cell trapped within the borders of another cell after having given up necessary cellular functions to survive independently outside of that cell? Would it be weird, or analogous to the human condition? Adaptation and collaboration become essential.

Using reductionism, the only point at which something close to a real-self can be identified is at the level of the genome. The body and cell surrounding the genetic code function to protect and propagate the code but are not the real-self. The genetic code appears to be responsible for biological forms and functions. How the code is expressed becomes the self in that moment.

In cellular life forms, DNA code appears to have permanently coopted biology for its own protection and propagation. This may not be dissimilar to how an RNA virus temporarily coopts a cell for its metabolic and reproductive purposes. It is not entirely clear whether DNA is functioning in an authoritative versus a cooperative endosymbiotic nature with the remainder of the cell. But what is clear, the expressed code determines many of the functions and appearances of cells—and humans.

To complicate this picture further, around 10% of the human genome appears to be of viral etiology. Retroviruses can incorporate themselves into DNA, such as human immunodeficiency virus (HIV), an RNA retrovirus that worms its way into human DNA, thus protecting and propagating its own code by this mechanism. In this case HIV exists in a chronic pathogenic or parasitic-type relationship to the host causing disease, disability, pain, and suffering. But much like bacteria became endosymbionts in the form of mitochondria to coexist within cells healthily and happily, other RNA retroviruses have incorporated themselves within the human genome in an endosymbiotic DNA form to the benefit of both the virus and the human. So, it is not unreasonable to say that human cells are archaea with endosymbiotic bacterial **and** viral descendants and remnants, which makes the real-self even squishier.

Genetic code seems to have combined, collaborated, and adapted—whether raw, viral, bacterial, or archaeal code—over billions of years to either highjack or form more advanced lifeforms to ensure the protection and propagation of their code. From single-cellular prokaryotes and eukaryotes to multi-cellular organism to differentiated-cellular organisms, the genetic code of life has found many ways to protect and propagate itself. Bacterial antibiotic resistance, plant photosynthesis, and the human form have a commonality in that although they are all different environmental responses and forms at their core these responses and forms further serve to protect and propagate the code of life.

When contemplating the genome as a potential real-self, or even a constant self, the dynamic nature of the code itself denies this constancy. It is entirely possible that what appears today as damage, error, or random mutations will be seen tomorrow as strategic nucleotide substitutions and transposing or jumping genes in response to the environment along a sliding scale of threat to safety. Indeed, the genome may be found to be much more plastic than imagined today and capable of reprogramming its own code. [37] The search for a static real-self will still be unfulfilled.

The modern Western cultural concept of the human self may be no more than a created ideation, cognitive construct, symbolic narrative, and held belief that does not reflect a biologic reality. Many traditional cultures do not disconnect the individual, band, tribe, air, water, and land from one another. They are viewed as one. Within this view, the apparent different parts do not and cannot exist nor function without the other. The self is confluent with the environment. Both are ever-changing within a cosmic flow. This vantage point seems much closer to biologic reality.

Here's something else to contemplate: the genetic code without its connection to the environment would appear inorganic, adynamic, and dormant, as no physiology would be produced. Life, the self, only exists within the context of the environment. The non-self, or the other, creates the self.

It is important to recognize that the concept of a self is an evolutionary survival trait. Although cellular life forms are tightly integrated into and reflective of the environment, there is separation from the environment—a boundary that defines and protects life forms and their internalized genetic code. Without the life form's awareness of the difference between its form and the environment, life as it is known today could not exist. The organic would not defend and would quickly revert to the inorganic without this awareness. In general, it is connection at the boundary that can enrich life, but it is encroachment of the boundary that can distress and even kill life.

A caveat: Life is being defined by the functions of a cell as previously described. There is the possibility that reproduction and propagation of code may exist outside of cellular life, outside of a boundary and its containments and functions. Crystals can reproduce. Peptides can congregate, propagate, and heal injury. Viruses are close to this idea of noncellular life in concept but require cellular mechanisms to reproduce and propagate their code. However, viroids and obelisks are capable of nucleotide based self-propagation without cellular functions. Prions are self-propagating amino acid, not nucleic acid, based infectious proteins that do not use RNA or DNA for their reproduction and propagation.

It looks possible that both amino acid and nucleic acid-based codes can be reproduced and propagated through noncellular energy sources and reactions. Is it also possible that something very close to cellular life could exist without a nucleic acid code to steer it? (see the nongenomic intelligence of the planaria.)—TBD.

For these purposes, the noncellular reproducers would not qualify as life as it is known, as the cellular life that has been defined previously. Under the TVST model, "life" is defined as a binary state, inorganic or organic. Something is either alive or not. However, in the future, perhaps, life will be viewed as lineages of propagating information, code, intelligence, and awareness that temporarily find themselves aggregated in a cellular form.

"The boundaries which divide life from death
are at best shadowy and vague.
Who shall say where one ends,
and where the other begins?"

Edgar Allan Poe

Multi-cellular and differentiated-cellular organisms have an even greater need to sense their composite colony of cells. These organisms require more advanced coding to protect themselves within an expanding systemic program. The individual cells within these greater colonies—multi-cellular organisms—have long since given up their ability to survive autonomously and independently. It doesn't make sense for the cells of multi-cellular and differentiated-cellular organisms to give up on the colony when survival without the specialized functions of the various colony members is not possible. Although, it appears cancer cells try infidelity to and autonomy from the whole, in general, this strategy doesn't work well in complex life forms. It culminates in poor results for both the cancer and the organism.

Higher evolutionary levels in complex organisms sense and coordinate the whole organism as a singular entity, something that the individual cells cannot sense and do on their own. There is a different sense of self at each level of evolutionary development and complexity. Perhaps the most notable, for good or bad, is the ability to construct and hold an image of the self at the human mental level. However, this constructed image typically does not depict the total reality and complexity of the human organism.

In many ways, the concept of the self is more of a reflection of the non-self, the environment or the other, integrated into biology and physiology for the protection of the code. The sensing of self versus other is tightly linked to a state of defense and a TR.

There may be implications here for therapeutics and the value of self-care, self-help, self-love, self-compassion, self-awareness, self-examination, self-realization, self-actualization, self-image, self-esteem, self-improvement, true-self, authentic-self, and identity reformation that all reenforce the focus on the self and, thus, may feed the defensive state and the TR.

The focus on the self, although necessary for surviving, may be physiologically counterproductive to healing, health, wellness, and thriving.

In fact, the concept of a self may only be truly necessary when in physical threat. Short of a state of physical threat, the focus on the self may be detrimental.

All threats push humans towards self-obsession. To simply try to stop being self-focused can make it more difficult to extinguish these self-protective strategies. Sometimes obsessive thoughts and compulsive behaviors become self-destructive; obsessive compulsive disorders are highly associated with increases in morbidity and mortality. Behaviors that make humans safe allow for the dissolution of the self and self-obsession. However, trying to stop being self-obsessed, ironically, and paradoxically, may increase threat physiology and make things worse.

Consider this: The connectotype in the brain called the *default mode network*, or DMN, is involved in self-reflection, self-contemplation, and self-evaluation. In addition, the DMN has concerns with what other people are thinking and social norms, and the associated contracts, ideations, constructs, judgements, narratives, and beliefs. Chronic threat-related DMN activity seems to be associated with self-protection, self-obsession, anxiety, avoidance, introversion, isolation, loneliness, rumination, and depression. This suggests self-attention can not only be driven by threat, but can be a *source* of threat, and is expressed in a threat phenotype and TR.[38]

Biologically, threat forces the necessity of the self, and ultimately, the isolation from the other. Safety allows for the dissolution of the self and integration with the other. Awareness of the self is necessary for the defense from threats, but chronic threat in humans causes anxiety, isolation, rumination, and obsession about the self. These amplify, perpetuate, and exacerbate the TR.

Life may neither be about finding one's self, nor about creating one's self. Life may be about losing one's self within a world of safety.

> *"For whosoever will save his life shall lose it,*
> *but whosoever will lose his life shall save it."*
>
> *Luke 9:24*

All awareness, memory, intellect, and ultimately individual consciousness, spring forth from the genome and its engagement and entanglement with the exposome.

Awareness of self allows for awareness of death. The human cognitive construct of a fixed and immortal self can give a sense of control, agency, meaning, and purpose to soothe existential angst. However, this formation seems to be a distortion, or little more than an illusion, of how humans engage and adjust to the environment, most notably at a cellular level.

The false-self is a defensive mechanism that ultimately births human's illusions of control within a world of little control. Illusions of control are some of the human species' most potent, pervasive, and perverse mechanisms for repressing physical and social existential angst. Humans' greatest and strongest fears are both a physical death and a social death. The knowledge of death and nonexistence is a curse of human neocortical awareness and intellect. However, the bigger curse from which humans suffer is the reflexic, unknowing, evasion of the aversive and painful feelings attached to the threat of physical or social nonexistence. In other words, repression. Both in awareness and unawareness, patterns of behavior are frequently oriented at trying to control and prevent physical, emotional, and social injury and annihilation. They can also be geared toward the avoidance of aversive feelings associated with these threats.[39,40]

"Suppressing the feeling only makes it harder to let them go. Expression is the opposite of depression."

Edith Eger

Illusions of control centered around ego, or false-self, protection, social viability, and the denial of mortality are analgesics for existential pain. One of the central human illusions of control is the construct of the static, perhaps permanent, false-self that fails to fully see the nature of an individual human's impermanence, as well as the interconnectedness of the greater species, with the world. In addition, this construct of a false-self fails to acknowledge the limits of human control within the greater world.

The human neocortex of the brain is distinct and unique in its capabilities of imagination and abstract thinking, constructing tools and technologies, along with advanced symbolic communications. The human neocortex is also capable of assembling narratives, holding beliefs, and creating an image of the self beyond the reality of the self. Within this context, these powers of the human neocortex can also become liabilities. The defense of the constructed imaginary false-self over the physical real-self and the genetic code can become problematic. It can be meeting the enemy only to find that person is you.

Modern culture exerts a sense of entitlement within the world. Humans have created a construct in a male protector God that has endowed superiority to the Homo sapien species. Through this construct, the world was provided to humans for human use (abuse). The God humans wanted is the one humans created and believe in. This belief in a false-God as well as a false-self is the surreptitious blindness that leads to much antisocial behavior in the world—evil. The great trickster—Devil, Satan, Lucifer—is at play. Belief, a human physiologic phenomenon constructed and stored within the human brain but lacking any objective truth, is central to not only religion but to neuroses, psychoses, and cults.

Humans have incorrectly surmised and believe that many other species lack "consciousness" and do not experience pain and suffering like humans do. This hubristic construct is not only wrong, but it has led to not just an apex predator status, but to the denial of the destructive nature of the human species.

Homo sapiens are uniquely aware of the possibility of their annihilation. Paradoxically, their associated illusion of control behaviors may be what ultimately leads to human annihilation. Compounding this paradox is the dilemma that modern Western culture and human social viability demand these hubristic constructs, narratives, and beliefs, and demand the protection and propagation of these constructs, narratives, and beliefs. This illusion of status and control helps to quell human existential angst. However, it also leads to enormous suffering.

An opioid for human neocortical awareness, awareness of being aware, and awareness of the possibility of annihilation, is the transformation of an ideation to a construct to a narrative to a belief, and then to an ideology. False maladaptive beliefs that become cultural norms can create escalating threat within the world. This type of threat is more significant in the world today, primarily due to its technological virulence and transmissibility. Threat is easily spread through modern medias. When "evil" is seen within the world, antisocial threat phenotypes in action, false belief and deception are usually running hand in hand with this evil.

The exchange made for the curse of human awareness and knowledge of human fragility may have been the ability to believe. Belief is not the equivalent of faith. Faith allows for not knowing, reconciling with the limits of our humanity, and letting go of control—or the illusions of control. Belief is founded on social contracts, created ideations, cultural constructs, and shared symbolic narratives. They may or may not reflect reality. Belief can perpetuate illusions of control and repression. Belief can help humans to survive acute threat, but when used chronically, belief frequently becomes maladaptive leading to increased conflict, disease, disability, pain, and suffering. Like an opioid belief is addicting, and if used as a chronic analgesic belief can take people, societies, and the globe down.

When illusions of control become maladaptive, there are many adverse consequences manifested in physical, emotional, behavioral, social, cultural, and spiritual illness, disease and disorder.[41] Having awareness of these illusions is essential to create safety within the world. Illusions of control are manifested primarily through social contracts, created ideations, cognitive constructs, symbolic narratives, and held beliefs, all functions of the distinctly human neocortex. Many if not most people are simmering and stewing in their repressions and illusions of control. This is a major source of conflict and suffering.

"When the norm is decency [safety],
other virtues can thrive: integrity, honesty,
compassion, kindness, and trust."

Raja Krishnamoorthi

Philosophy contains the constructs of ethics, morality, virtue, meaning, purpose, and power as aspirations and goals to have a life well lived. Philosophy paints an image of the self to strive for, yet anything less becomes a threat to that image of the self. The associated TR thus becomes a threat to the real-self. Like a photograph or painting once finished, this false image of the self is stagnant and only weathers with time. Humans tend to compensate for this tarnishing with self-righteous victim-martyr-hero narratives that serve to polish the false image. In fact, humans in threat participate in incessant, mindless, non-conscious competition within symbolic victim-martyr-hero narratives. In addition, threat states induce catastrophizing, an emotional and mental amplification of predicted or perceived events, that easily get spun into narratives and beliefs then woven into our victim-martyr-hero stories for social leverage that in chronicity exhaust social connections. Humans are frequently just failed actors within these amplified and embellished stories missing out on living their real lives and real connections.

Ethics, morality, virtue, meaning, and purpose are constructs of the human mind unnecessary to prosocial behaviors, health, or happiness—see bacteria and bumblebees. Safety and safety physiology lead to a prosocial and meaningful life. The biology of safety feels good and creates good without the baggage of constructs, narratives, and beliefs.

The real-self is relative, contextual, and forever changing. When in a threat phenotype and physiology virtue, meaning, and purpose are frequently just created ideations, cognitive constructs, and symbolic narratives, yet they are the real products of a phenotype and physiology of safety—they are naturally present, not imagined, within safety. They are not aspirational stories nor lofty goals of a human in threat but arise spontaneously from the physiologic soup bathing the real-self when safe...only to fade if safety recedes.

"The greatest of evil included all human motives in one giant paradox. Good and bad were so inextricably mixed we couldn't make them out; bad seemed to lead to good and good motives to bad. The paradox is that evil comes from [hu]man's urge to heroic victory over evil."

Ernest Becker

The Cortex and Consciousness

"I think, therefore I am."

Rene' DesCartes

"I sense, therefore I am."

Ali F. Oeus

Beyond the mechanics of cellular and cortical awareness lies the idea of consciousness. The Western concept of consciousness is incomplete. It primarily refers to cortical awareness—a process that happens when awake—as being the extent of consciousness. Notably, Eastern concepts of consciousness hold that all that exists is consciousness i.e. consciousness is a universal constant. This may be closer to reality. What consciousness is precisely and where it comes from has remained elusive. It is clear, however, that the use of awakeness or thought to define consciousness is inadequate.[42]

It is fair to say that consciousness is, at least in part, the sensing, awareness, and perhaps the intention, of the self and being alive. In this regard, consciousness may have preceded or co-evolved with the boundary, or cell membrane, as a prerequisite for life as it is known today. The sensing and awareness of the self in contrast to the other is required for the metabolic, defensive, and reproductive processes of an organism. Therefore, all life forms have consciousness. To be conscious does not require a nervous system, brain, cognition, or a mind—whatever a mind might be.

So, how much biomass is required for consciousness? What about something as simple as a virus or even the simple genetic code of a virus—is there consciousness at this level?

A virus is not classic cellular life such as an archaea, bacteria, or human. A virus is little more than a protein shell containing genetic code and does not have classic cellular structures. Nor does a virus possess intracellular metabolic and reproductive networks. A virus relies on invading cells and hijacking metabolic and reproductive networks to sustain and propagate its code. However, in hijacking another, it is clear a virus can distinguish itself from the other. Arguably, a virus understands self from other and therefore has consciousness; without consciousness, self-awareness, and knowledge, its genetic line would cease to exist. Something with consciousness that reproduces, yet without the trappings of defined cellular life, blurs the lines as to what life really is. This suggests that consciousness may not be found within a brain, tissue, or even a cell, but at a molecular or even energetic level that has yet to be defined.

Is consciousness an energy not yet explained by the classical sciences—physics, chemistry, biology—and something that exists in the nonclassical quantum world? These are tough questions to answer. But suffice it to say, all cells have consciousness and maintain their individual consciousness. This is still the case when the cells belong to multi-cellular and differentiated-cellular organisms and are a part of their higher forms of consciousness.

Multi-cellular organisms require consciousness of the whole, not just consciousness of the individual cells. This multi-cellular consciousness is a more complex awareness than its single-cellular counterpart. Multi-cellular consciousness involves integrating, coordinating, and synchronizing many cells, all devoted to the protection and propagation of the whole organism, the species, the code.

Multi-cellular consciousness differs from single-cellular consciousness in that it includes the awareness of the whole. Advanced, complex, and larger organisms that have differentiated cells, nervous systems, and brains—differentiated-cellular, multi-tissue, and multi-organ organisms—are also highly dependent on all the cells within the organism for survival. These brain-containing organisms, such as humans, have an even more complex consciousness that include awareness of the greater self. Yet, this awareness of the greater self does not contain an awareness of the individual cells that make up the whole of the colony—the small selves that make up the big self. In fact, the vast majority of the greater self operates out of the awareness of the brain unless threatened or injured.

In addition, perhaps uniquely, humans have awareness that they are aware. This unique level of awareness likely correlates with the unique neocortex of the human brain. Not only does the human awareness of being aware reside within the neocortex of the human brain, but so do human social contracts, created ideations, cognitive constructs, symbolic narratives, and held beliefs. Perhaps, this unique level of awareness can be best characterized by the term "sapioconsciousness" as advanced Homo sapien neocortical processes seem to be the center piece to this awareness and knowledge, as well as, to the formation of these contracts, ideations, constructs, narratives, and beliefs.

Also, it is within the neocortex of the human brain that an imaginary static construct of the self is created with an associated rigid narrative and fixed belief attached to the constructed self. The creation of a false-self or an image of the self, can become problematic. The real-self is ever changing and illusive. Whereas the constructed false-self can be an illusionary ideation, construct, narrative, and belief of some inflexibility and rigidity and, tending towards stasis.

The false-self can be confused with the real-self, but it should not be. The real-self is to be defended as needed. The false-self can be a temporary construct for the camouflage and defense of the real-self when under acute threat. However, it is not an accurate representation of reality. It does not represent the multiplicity of reality, and the real-self, the multiplicity of selves. When chronic protection and propagation of the false-self usurp the protection and propagation of the real-self, a state of nonconsciousness can result in dysfunction and disease. The classic example of this is the story of Narcissus who did not fall in love with himself, but with the *image* of himself.[43]

> *"Here is a face looking at a face, and the problem*
> *is the image of the thing is never actually the thing.*
> *You try and grab it and it's not there. It disappears."*
>
> *Jane Alison*

When human cortical brain functions are dramatically turned down or turned off, humans can experience cortical unconsciousness or loss of cortical awareness, such as with safety associated sleep or threat associated shock. These states occur when the reticular activating network, ascending from the brainstem, is downregulated or damaged. Unconsciousness states come from the bottom upward. Notably, in these states, multi-cellular and single-cellular consciousness still exist.

Alternatively, under acute threat social connections, contemplative ideations, complex cognitions, symbolic narrations, and declarative memories are not of value in this tiger fight. These Homo sapien neocortical functions go offline as more primitive brain functions are prioritized to defend the real-self. There is a relative loss of "sapioconsciousness" in this state, yet awakeness persists.

If chronic threat persists, these brain areas are chronically downregulated or dissociated. They may also be deconstructed, or even destructed.

Chronic social threat presents with another unique phenotype. Here, humans can also experience less than full consciousness or a state of nonconsciousness (not unconsciousness) when the protection and propagation of the false-self takes priority. This dissociation from the real-self is a nonconscious state marked by illusions of control, emotional repression, behavioral constraint, thought suppression, and even paradoxical threat biased thought obsession.

The neocortical functions that are offline in acute physical threat states within chronic social threat states are preoccupied with social and cultural norms and scripts, and the compliance with these norms and scripts to avoid emotional and social injury, while hoping to afford safety. While the combined neocortex searches embedded scripts for what is proper, the medial prefrontal cortex inhibits emotional and behavioral striatal flow. Thus, the medial prefrontal cortex can inhibit cortical activation from the transmissions of the ascending striatal network. This, in turn, creates the physiologic mechanisms of disconnection, constraint, repression, and suppression. In chronic social threat states, full consciousness is diminished as the physiology and drives that bubble up from below are held at bay from sensation, action, and awareness.

But also, the distraction and preoccupation of the neocortex in trying to protect and project the false-self thus preventing the experience of connection with the physiology and emotions from below, prevents the connections to all the other of the world.

This is a place of being stuck in the scanning of social and cultural contracts, ideations, constructs, narratives, and beliefs, while constraining, repressing, and suppressing the physiologic flow from below—disconnected from the rest of the full reality of the exposome—interosome and exterosome—to not only grind on and grind down the neocortex, but to be trapped in the neocortex. To be trapped in one's mind is to suffer.

Notably, this process is very different than the complete disinhibition of emotions, behaviors, and thoughts and the relative neocortical quiescence that may be seen in some more acute threat states. This losing one's mind can also lead to suffering.

These nonconscious phenotypes come both from the top downward and the bottom upward. In these relative nonconsciousness states, multi-cellular and single-cellular consciousness persist.

There are many phenotypes of threat. However, in general, in a state of threat, humans are more focused on survival and the self—real or false—than connection and the other. In threat, the world becomes very small as the self becomes very large. By contrast, in safety, the world is very large, and the self becomes very small. Realities become relative to threat versus safety. When in social threat, humans are particularly focused on the image of the constructed false-self. This strategy can be adaptive and effective in acute threat states, but the strategy can become maladaptive and detrimental in chronic threat states.

Recognition of physiology within cortical consciousness is how to realize emotions. Emotions are physiology playing out in the brain. They become conscious emotions, or feelings, when this physiology reaches the cortex of the brain and is acknowledged. Raw emotions are generated at the top of the brainstem in the mid brain prior to reaching other subcortical structures, such as the limbic network, or subsequently cortical awareness. Raw physiology, subcortical emotions, and cortical feelings are all designed to get us to do something, frequently something different than what is presently being done. Top-down neocortical functions can prevent this physiologic drive from happening. But at times, neocortical inhibition is too slow and too late; the impulse has already taken hold and directed action.

In humans, cortical consciousness in the form of acknowledged feelings is the representation and expression of what the cells are experiencing. Being fully conscious of emotions at a cortical level and being aware of this physiologic reality, and safely acknowledging, accepting, and expressing these feelings is required for the integration, coordination, and synchronization of the conscious experience across the entire organism—from the cells to the cortex. This appears to be a requirement to keep humans healthy. Anything that blocks this process may lead to disease, disability, pain, and suffering.

Contemplate this: for a human to be fully conscious is to experience and accept reality. This reality includes the beauty of human connectedness, if not confluence, with the other. It also encompasses the stark reality of human frailty and vulnerability, and how very little humans can know and control. Solace can be found in the fact that outside of thinking, speaking, doing, and controlling is a space of comfort, connection, love, and awe. This space, a deeper and fuller consciousness, is profoundly healing.[44]

The awareness of the connectedness of self to the other is most available when humans are safe, and when a defense is unnecessary.

It is daunting to think that the human body is made up of just under 40 trillion cells each inhabited by thousands of endosymbiotic mitochondria, colonized by just under another 40 trillion exosymbiotic microbes. Within this massive colony of mitochondria and cells, each individual mitochondrion and cell are capable of local awareness, intellect, and control. Only 16 billion of these cells are cortical brain neurons. Less than a billion of those neurons are devoted to cortical awareness, intellect, and control. Given the numbers, it seems improbable that the totality of the mind, the self, and consciousness lie solely behind the eyes and between the ears in a relatively few cells within the human brain.

Most of the human experience occurs outside of the cortex of the brain, and outside of cortical awareness, knowledge, and control. Additionally, a third of the human experience is in a state of sleep. Even when not awake, the brain's connectotype is ever changing. There are always various cortical hubs, networks, and functions coming online and offline, depending on the stage of awareness, awakeness, or sleep, paralleling the state of threat versus safety.

All this argues that when trying to cure disease, disability, pain, and suffering, working at the level of the cell and consciousness is where to start. Employing systemic and diffuse strategies that extend beyond the imaginary constructed borders of the mind, body, and brain. This requires a truly comprehensive approach. Cognitive awareness from education and counsel, in conjunction with pills and procedures, are all still valuable tools in health and disease care. In the end, however, they are ultimately inadequate for achieving healing, health, wellness, and thriving.

In fact, TVST suspects that at times, the contracts, ideations, constructs, narratives, and beliefs of the Homo sapien neocortex—the constructed false-self in particular—may be the causes of a sustained TR. Thus, these functions can also be a barrier to initiating a SR. These false contracts, ideations, constructs, narratives, and beliefs can propagate disease, disability, pain, and suffering. They can prevent healing, health, wellness, and thriving. Arguably, their deconstruction may be of benefit in initiating a SR and healing. This may explain some of the promising results being seen with the use of psychedelics in treating a variety of threat-related disorders.[45]

"You're only as young as the last time you changed your mind."

Timothy Leary

The psychedelic's mechanisms of action are variable and debatable in terms of precisely how they work. The popular theory has them working at various monoamine and cytokine cell receptor sites to cause cortical dissociation and hallucinations, and even neuroplasticity and neural growth.[46] Another possibility is that they are various forms of toxins, a foreign compound, that through a TR, cause cortical dissolution, dissociation, and hallucinations. It may be that some are a receptor signaler, and others, a toxin.

Another mechanism of action for their potential healing properties could be through modifying epigenetic coding. This could explain their effectiveness, robustness, and duration of action. For example, the epigenetic changes associated with cannabis use suggest a bias towards threat coding, but more research is needed to clarify this initial impression.[47] Some psychedelics may initiate the freedom from the inhibiting, managing, and governing oversight of some neocortical structures, freedom from the default mode network and self-examination, and freedom from defensive states through the induction of safety physiology to light up and strengthen a deeper connection and resonance with both the internal and external worlds.

Psychedelic effects have been reported as profoundly enlightening and healing, and at times, curative for mental health disorders. Given the durability of the response, such as with psylocibin, one must seriously consider if some of these psychedelics might not be working at the epigenetic level and affecting threat versus safety signaling and bias. If this is the case, the brain-centric theory for the mechanism of healing from psychedelics would be replaced with an epigenomic and systemic theory of healing through safety signaling. As a result, every cell, not just the neurons, would be implicated in the response.

Notably, the Homo sapien neocortex undergoes dissociation in both threat and safety—shock versus sleep are examples. Dissociation is typically thought of as pathological, but technically some part(s) of the brain is always dissociated. The dissociation of threat is decidedly different than the dissociation of safety. Both shock and sleep are profoundly dissociative states on opposite ends of the threat-safety spectrum.

Schizophrenia is also profoundly dissociative and marked by disconnection within and atrophy of the phylogenetically newer cortical structures of the brain. Perhaps most notable is the dissociation of the ventromedial parietal lobe (precuneus) that functions to integrate the world with the self. This level of dissociation can lead to a disconnection from reality in the forms of delusions and hallucinations as older, deeper brain structures operate more autonomously from the higher areas of sensory integration.

Alternatively, awe seems to be dissociation primarily from the regulatory and governing prefrontal brain functions, thus allowing profound activation and connectivity throughout the rest of the brain. Some psychedelics tend to mirror awe in their effect much more than schizophrenia. Clarification of this effect will be helpful in determining therapeutic uses of the different psychedelics.

Moving forward, it seems prudent to evaluate the psychedelics with an eye both on threat and safety biomarkers. This differentiation would help weed out the psychedelics that are operating as toxins, a threat, versus others that mimic safety signaling to promote healing, plasticity, growth, and resiliency.[48]

Experiences in nature, being inspired by moral behaviors, being moved by music, or spiritual contemplation, and some psychedelics, all seem to promote mental and physical health. This is attained through invoking the physiology and connectotype of awe. Awe is an emotion often considered ineffable, indescribable, and beyond measurement. Awe shifts neurophysiology, likely all physiology. In addition, it diminishes the focus on the self, increases prosocial relationality and greater social integration, and induces a heightened sense of meaning and well-being. Experiences of awe that arise from nature, music, collective movement, spiritual experiences, and psychedelics add to resiliency. Safety makes awe, a higher-level state of consciousness, accessible.[49]

Full consciousness is profound awareness and connection that seems to go beyond thoughts, contracts, constructs, narratives, beliefs, and intellect, and even beyond cellular, subcortical, and cortical awareness—something really big. Arguably, cellular consciousness arises at the genome's interface with the exposome, and thus, cannot be separate from the environment. Consciousness appears to be systemic and diffuse. Consciousness, as it relates to the other, may be a form of energy extending beyond physical borders. Consciousness may be related to the nonclassical mysterious quantum world of energies and entanglements.[50] Contemplate the notion that if consciousness could evolve from single-cell organisms to multi-cell organisms, to differentiated-cell organisms, it may be more universal than humans can perceive.

"The self dissolves while connection and consciousness expand within the depths of safety."

Ali F. Oeus

In summary, consciousness is not unique to humans, contrary to claims. This claim, in fact, conflicts with evolutionary principles. Consciousness is at least as old as the biotic world, and perhaps as old as the prebiotic world and the universe itself. It seems to be a requirement, if not prerequisite, for the establishment of the fundamental parameters for life—a code and boundary, along with metabolic, defensive, and reproductive functions.

Consciousness evolved with the evolution of life and is more complex and layered in more complex and larger life-forms. Perhaps consciousness is most complex within humans. Human consciousness includes basic sentience through cortical awareness, neocortical awareness, and "sapiocortical" awareness—including the awareness of *being* aware.

Existential awareness is most acute in the human species as well. The knowledge of our pending potential annihilation is frightening. Other threats heighten existential angst even further. Humans tend to soothe this painful possibility of non-existence with illusionary constructs of immortality, false-self formations, and the relative nonconscious state. The nonconscious state is an opioid for existential angst. It protects and calms the system in the short run but runs the risk of addiction and destruction when used over the long run.

So, what is the mind? An ideation and construct of consciousness with agency? It is very hard to define, and perhaps the term should be extinguished. However, it may be fair to say that the human "mind" comes from the level of sentient awareness, but best aligns with the Homo sapien neocortex where advanced social connections, created ideations, complex constructs, executive functions, symbolic communications, and factual memory reside. It is certainly this part of the human brain that accounts for some of the brilliance of the species that is seen in technology, literature, art, and music. But the human mind is a double-edged sword, and the human neocortex can also lead to bad contracts, paranoid ideations, frail constructs, toxic narratives, and false beliefs. In addition, the neocortex can produce the false-self that interferes with feeling, protecting, and nurturing the real-self, others, and the species.

"I was taught the human brain was
the crowning glory of evolution so far,
but I think it's a poor scheme for survival."

Kurt Vonnegut

It has been said humans experience the curse of consciousness. If this is the case, then all life has been given this curse, not just humans. It may be more accurate to state that humans have been given the curse of the Homo sapien neocortex—"sapioconsciousness"—and this has become a curse to all life. The Homo sapien neocortex cuts both ways.

Uniquely Human

What is different about Homo sapiens? In general humans have little fur, but so do other species. Humans have opposable thumbs, but so do other species. Humans walk upright and somewhat awkwardly, but again, so do other species. Humans are relatively slow and weak, but there are other species that have survived with these traits. Being upright resulted in hiding female genitals. Perhaps the evolutionary compensation for this was a subconscious symbolic adaptation of uniquely extroverted lips and adult females having unique relatively permanent enlargement of their breasts. Humans uniquely wear clothes, shoes, and ornaments to protect, hide, signal, or create something. But what is really unique in Homo sapiens is their brains, specifically their new brain surface or neocortex.

All mammals have a neocortex for sensing and processing information from the interosome and exterosome and then the neocortex responds to this information with a variety of outformations. However, Homo sapiens' neocortex is massive.

Even relative to other great apes, including other extinct human ancestors, the ratio of the Homo sapien neocortex to other brain structures is twice as large. Only Neanderthals' brains compared. Most of the rest of the Homo sapien brain is very common with other mammalian species.

The Homo sapien neocortex, perched above the nose and over the eyes, with prominent temporal and frontal lobes, accounts for modern human's large and awkward heads and long and wide vertical foreheads. From an evolutionary standpoint this is the newest expansion of the neocortex occurring only 300,000-250,000 years ago. The Homo sapien neocortex consumes 80-90% of the total energy the brain consumes. It is a metabolic sink which has significant implications in different defense and survival strategies.

The Homo sapien neocortex is so unique it rightly deserves its own descriptor. Something like "sapiocortex" may be appropriate; it is truly the one thing that distinguishes Homo sapiens from all other species. "Sapio" is based on the Latin verb "sapere," which means "to be wise" or "to have sense." When naming the species, it is assumed this was not sarcasm but an optimistic hope for the species. Although, not universally wise, Homo sapiens are profoundly smart, and perhaps that is where the confusion lies. Homo sapiens are different from any other species in both their neocortical aptitudes and ineptitudes.

This "sapiocortex" is what gives Homo sapiens the ability to form complex social contracts, vivid ideations, cognitive constructs, and symbolic narratives. The "sapiocortex" has enabled humans to connect, dream, create, communicate, and dominate the Earth as the apex species.

When Homo sapiens feel safe, the "sapiocortex" allows for empathy, generosity, compassion, and kindness. In addition, it can create spectacular works of music and art, advanced tools and technologies, and transcending poetry and prose. Within safety, Homo sapiens can flourish and thrive in peace.

Threat physiology biases the "sapiocortex" towards defensive phenotypes and strategies. Selfishness and self-protection are prioritized. Reflection gives way to reflex. Contemplation gives way to reaction. Imagination gives way to paranoid ideation. Creation gives way to vigilance. Transcendence gives way to rumination and obsession. Faith and peace are replaced with rigid belief and conflict.

Unfortunately, Homo sapiens live in a modern world of chronic unrelenting threat. Individual behaviors are biased to be antisocial or asocial, social contracts divide into us versus them, paranoid ideations vilify the them, symbolic narratives destroy the others, and rigid beliefs entrap in suffering. These threat phenotypes of Homo sapiens demonstrate the struggle for human survival, which has turned the species into the apex predator on Earth.

The Homo sapien brain is to be marveled at in its capacity for good as well as to be scrutinized in its capacity for evil—all along the spectrum from safety to threat. Wisdom isn't so much learned, as it is a state of safety. Wisdom, great wisdom, is always within Homo sapien DNA, but only to be expressed when in safety. Therefore, to find wisdom it is advisable to seek and create safety.

Intelligence and knowledge are not the equivalent of wisdom, and they are not needed to be wise. Human cortical intelligence and knowledge as they are typically described refer to "sapiocortical" intelligence and knowledge. Whereas there is a diffuse level of nongenomic, genomic, mitochondria, cellular, tissue, organ, network—systemic—intelligence and knowledge. It is within the integration of the "sapiocortex" and the rest of the system where wisdom can be found, an ultraconsciousness, that frees the mind from the dominance, and at times stupidity, of maladaptive "sapiocortical" contracts, ideations, constructs, narratives, and beliefs.

The "sapiocortex" is a vastly complex connection of circuitry and modifiers of the circuitry but here it will be grossly simplified for concept. A good place to start in understanding the "sapiocortex" is where Homo sapiens seem most unique in function related to other species.

First, Homo sapiens have a profound need for social connection, cooperation, collaboration, and adaptation to survive within the natural world. This is not to deny the social nature of every cell and every other organism, but simply to acknowledge that humans as large, differentiated-cellular, weak, slow, and somewhat awkward beings really need each other to survive. The ventral medial prefrontal cortex, the expanse of human brains just above the eyes, is where advanced social connection and social contracts are best localized.

In the precivilization world of the hunter, forager, and gatherer, survival required not just the larger group but the coordination of the group. Communication became key in this regard leading to an advanced symbolic communication network—language. The human communication network includes the rudiments of more primitive communications including vocal grunting, groaning, yelling, screaming, laughing, and crying, all inclusive of associated facial expressions and body postures. The network also includes the use of symbolic gestural, verbal, along with written information and communication—the likes of which other species do not compare. The lateral cortices, the expanse of human brains above the ears is where advanced language and narratives are best localized.

The symbolic communication network allows for the exchange of information throughout the species, but this demands an advanced server to store this information. Declarative memory, historical and factual semantic memory, is best localized to the servers of the medial temporal lobes, the expanse of brain folded towards the center between Homo sapiens' ears.

The Homo sapien brain is also wildly imaginative. Case in point: symbolic communication. Information from multiple brain centers can be pooled and played with to manipulate the past, enhance the present, and anticipate the future, and move the abstract to the concrete. These functions best localize to the ventral lateral prefrontal cortex approximate to being inside of the temples.

The human brain has the unique ability to take multiple inputs, informations, and create internal constructs prior to generating multiple outputs, outformations, and external constructs. This process has resulted in the technological evolution of tools, such as knives and wheels, all the way to rockets and the James Webb telescope. The dorsal lateral prefrontal cortex, the expanse of brain behind the top of Homo sapiens' massive foreheads, is where advanced cognitive constructs are best localized. This is where human thoughts come to rest, sometimes stick, and sometimes stick too long. This is also where executive functions such as working memory, reasoning, planning, strategizing, problem solving, judgement, and mental flexibility, and are best localized.

In contrast to these advanced brain functions, emotional, visceral, and even fascial feelings arise to awareness from below in the more primitive, deeper, and central brain structures best localizing to the midbrain, and the cingulate (girdled structure in the middle of the brain) and insular (buried deep and insulated within the folds of the brain) cortices. Information from below in the forms of fascial, muscle, heart, and gut inputs arise to drive certain emotions, behaviors, and thoughts. This information eventually reaches cortical realizations in the forms of feelings, impulses, and intuitions, all primarily designed to protect and propagate the human code. The self does not actually exist within these structures. However, a sense of self does seem to reside within the more primitive cingulate and insular cortical structures in very close proximity to visceral and emotional feelings.

(Interestingly, although the insula is felt to be an older cortical structure, the anterior portion of the insula seems to be regulated similarly to the uniquely human neocortical structures of the brain. This area may represent a newer structure within the insula that serves as the receptacle, if not the womb, for beliefs, another unique function of the human brain.)

Information and physiologic flow from above, from the "sapiocortex," many based on social contracts, created ideations, cognitive constructs, and symbolic narratives, also drain to and are stored in these lower cortical structures. This input's output can be belief. In other words, a subjective sense and feeling that something 'is' without any objective information to support that outformation—visceral and emotional feelings connected to a contract, ideation, construct, and narrative created solely in the human "sapiocortex" manifested as a belief within these deeper and older cortical structures sit alongside "gut" feelings, instincts, intuitions, and the sense of self. The "sapiocortex" in some sense has highjacked these older cortical structures for the storage of beliefs—well below and hidden from the executive functions of reason and judgement.

General "Sapiocortical" Functions

- *Dorsomedial prefrontal cortex (dmPFC)*
  – emotional, behavioral, social regulation
- *Ventromedial prefrontal cortex (vmPFC)*
  – social contracts
- *Ventrolateral prefrontal cortex (vlPFC)*
  – ideations and imaginations
- *Ventrolateral prefrontal cortex (vlPFC/frontal language cortex – Broca's Area)*
  – symbolic narratives
- *Dorsolateral prefrontal cortex (dlPFC)*
  – cognitive constructs, executive functions
- *Medial temporal cortex (mTC)*
  – declarative memory
- *Ventromedial parietal cortex (vmPC-precuneus)*
  – integration of the external and internal environments with each other and the self
  – imagery and reflection on the self
- *Anterior insular cortex (aIC)*
  – embodiment of emotions
  – sense of self
  – the womb for held beliefs – below rational thought

(The older insular cortex houses interoception – emotional and visceral feelings – "gut feelings".)

Interestingly, humans with social anxiety have diminished brain volumes in areas of the PFC, correlating with poorer cognitive function and socioemotional engagement. Yet they also have increased CC and IC volumes consistent with amplified visceral and emotional feelings, and a focus on self in relation to other humans, as well as findings of morphologic changes at the top tip of the striatum in the nucleus accumbens consistent with inhibitory regulation.

It is helpful to visualize a flow of input as information coming into the brainstem from the interosome and exterosome, arising to the level of the midbrain for initial subcortical, subconscious, reflexive projection up the striatum, to then arrive at the cortical brain for acceptance, modification, or inhibition. In Homo sapiens, the "sapiocortex" with its social contracts, created ideations, cognitive constructs, symbolic narratives and held beliefs gets to weigh in on this input of information to accept, modify, or inhibit prior to outformation and output.

Good input leads to good output. Bad input leads to bad output. In many cases biased, prejudiced, bad, faulty, or false contracts, ideations, constructs, and narratives lead to biased, prejudiced, bad, faulty, or false beliefs. A problem arises when humans are under threat in that neocortical control is relatively diminished and biased towards TRs, while the paleocortical functions are preserved (at least for a while). Belief becomes primary over reason. Biased, prejudiced, bad, faulty, or false belief becomes maladaptive and potentially destructive physiology leading to physical, emotional, behavioral, social, mental, societal, and spiritual disease, disability, pain, and suffering.

Humans are of infinite minds and the operations of these minds can occur in conjunction or in opposition to each other. This cognitive-belief dissonance between the mind of the neocortex and the mind of the paleocortex registers as conflict or error in the anterior cingulate to subsequently activate or deepen a TR and its sequalae. In addition, the more we invest in paleocortical associated belief the harder it is to activate "sapiocortical" cognition. Reasoning with believers and cult members will fail until they are made to feel safe. Alternatively, if under constant and chronic threat humans will believe almost anything.

"He's out of his mind.
That's what he is out of."

Jacques Clouseau – Pink Panther

An important concept to revisit is that the purpose of life is to continue life—the purpose of life is to protect and propagate the human genetic code. Everything beyond that is contract, ideation, construct, narrative, or belief. When humans get stuck in chronic threat and the associated "sapiocortical" processes that make humans so unique and arguably the apex species/predator, then humans can either present with being out of their minds/"sapiocortex" (T1 phenotype) or as being stuck in their minds/"sapiocortex" (T2 phenotype) In these states, humans can deviate from the primary and perhaps sole purpose of life…to everyone's suffering.

The "sapiocortical" brain is where humans are aware that they *are* aware, and aware that they will die and disintegrate. Belief is not a salve, but an opioid for this chronic wound. Belief can make it feel better, but it will only prevent its healing. In this regard, it may not be the unconstrained primitive brain, but the unconstrained modern human brain that creates the platform for most of human suffering. Threat- based contracts, ideations, constructs, narratives, and beliefs may have their own feedback loop in that threat turns down "good" "sapiocortical" functions and "bad" "sapiocortical" chronic threat algorithms feedback to turn down, deconstruct, or even destroy the "sapiocortex."

However, this doesn't leave humans to appear more like

healthy integrated primitive organisms, but more like autonomous cancer cells. The solution remains not just in fixing the threat-based contracts, ideations, constructs, narratives, and beliefs, but in true safety to maximally, protect, propagate, and potentiate the human species. It is within this space and time of safety that humans transcend the "sapiocortex" and themselves to find deeper connections, to alleviate fear, dissipate the need for belief, and find faith…to heal, be healthy, and be happy.

"To recognize one's own insanity is, of course,
the arising of sanity, the beginning of
healing and transcendence."

Eckhart Tolle

It is important to note here the concept of "last in—first out." The sapiocortex is from an evolutionary standpoint the last in and in states of threat it is the first out. Through a cascade of physiologic mechanisms that include mitochondrial DNA insertions into the genome of "sapiocortical" cells and other intrinsic genomic alterations, to DNA/RNA, purine, glutamate, monoamine, peptide, and hormone signaling, to changes in blood and electrical currents threat deactivates, dissociates, and eventually degenerates the "sapiocortex". From and evolutionary perspective the "sapiocortical" functions are of little value in pathogen and predator attacks. Therefore, the "sapiocortex" is very vulnerable in threat states. These mechanisms of threat associated dissolution have profound implications for neurodegenerative and neuropsychological illnesses and diseases, and even for some forms of cancer.

Determinism

The hypothesis of determinism has its foundation in physics. It states that all events, including human action, are ultimately determined by a set of coordinates present at the very moment of the creation of the universe that then determine all events to follow.

Philosophers and religions have also dabbled in the concept of determinism. This kind of determinism extends beyond just the world of mathematics and physics to consider the roles of humans, meaning within the world, and accountability of action within a predetermined world versus a world of free will. With regards to religion, the roles of a superpower(s) and God(s) are at play.

Determinism within philosophy states that any decision or action of a human being is predetermined. For a human to make any decision other than the one that they make is impossible. Whereas, within the constructs of religion, a superpower(s) or God(s) dictate and determine all actions—a master puppeteer so to speak.

Within TVST, it is postulated that it is primarily the exterosome engaging the genome that determines the phenome, and physiology, thus determining metabolism, emotions, behaviors, thoughts, and connections. One can see deterministic principles involved here and the implication that the exterosome has control, not human beings, over these functions, or at least more control than humans perceive or believe—perhaps a lot more.

If the exterosome is a derivative of the physical world and if this world is deterministic then, yes, free will is limited, but is it absent? So how deterministic is the physical world, really?

It is worth taking a moment here to simplistically examine the physical world from its inception. How old the universe has remained unclear, but it is older than 14 billion years in age, perhaps much older. The observable universe has an approximate diameter of 100 billion light years and is continuing to expand, maybe even accelerate in its expansion. It contains at least 200 billion galaxies, each with hundreds of billions of stars. The universe is big, perhaps infinite, but perhaps not. The universe appears to have come from the explosion and expanse of a singularity—a single point where radiations are compressed and matter is extremely (infinitely?) dense, and mass and energy are extremely high—The Big Bang.

In physics, energy is defined by both the acceleration of

mass and the frequency and wavelength of radiation. The explosion of a singularity displaces both matter and radiation from its core and leads to the dance perceived as the universe. This explosive force goes out in all directions and dimensions, matter and radiation in motion, to create the expanse of space and time. The physical world is not a static state, but a state of constant action and reaction. Stasis would mean the collapse of this world of space-time, and so would the loss of the relationship to the other mean the collapse of space-time. Motion within the relationship to the other create these dimensions. No motion, no other, no space-time.

"All is flux, nothing is stationary."

Heraclitus

It has been theorized that some particulate forms of radiation are without mass—electrons, photons—but it is looking more likely that their mass is just extremely small. In general, particulate matter does have mass, and it is this matter that deforms the four-dimensional fabric of space-time to manifest gravity. Gravity, in turn, holds all matter in relation to each other. Gravity then bends and warps the waveforms of radiation. This combination of accelerated matter and plastic radiation in action and reaction results in the compressions, expansions, vibrations, spins, currents, swirls, eddies and orbits of the cosmos within the constraints of space, time, and gravity that ultimately define the universe.

Although a daunting task, if all coordinates were known at the very moment of the inception of the universe, then perhaps all events to follow would be calculable and predetermined. However, this scenario presumes that there are not outside or other forces at play—i.e., the world beyond this universe is absent or at least sterile and impotent in action and influence. It also dismisses the possibility of randomness.

The quantum world suggests that there is an element of disorder, with possibilities or probabilities, rather than order with certainty and precision (thus detracting from determinism). This is reflected within the very very small entities of the quantum world that is filled with uncertainties—where energy transitions from quanta to spins to packets to particles to strings to waves to appearances and disappearances in random locations, yet, always in relationship to, if not entanglement with, each other and the whole.

Of greater concern is that the world of the mathematician and physicist seems to have omitted the world of the biologist, likely assuming that all biology is governed by the laws of math and physics, the laws of the physical world. If this is not the case, and biology is not strictly governed by mathematics and physics, then absolute determinism as defined is less likely. At some level free will, or choice, is a possibility.

The mathematician and physicist's calculations seem to be missing an undefined but important force—the force of life. Life takes the reactivity of matter and radiation and organizes these substrates and energies into order and purpose—to protect and propagate itself. This is the fundamental code of life.

Life separates the organic from the inorganic in a dance within a dance. Life seems to be held together by consciousness, much like the relationship between matter and the field of gravity, the two in observation are inseparable. The inorganic integrates the field of gravity to hold the universe together in its dance, while the organic co-ops the field of consciousness to hold life together in its dance within the dance. The four dimensions of space-time are warped by the influences of mass, energy, and gravity while lifeforms and consciousness manipulate mass and energy—suspended under a twilight canopy life alters the clouds and the stars within this fifth dimensional dance.

As with gravity, consciousness may be a field—and not a force per se. Gravity is manifested and apparent through the observations and behaviors of inorganic matter and radiation. Matter manifests gravity, and gravity holds matter in relationships even to be pressed into singularities.

Consciousness is manifested and apparent through the observations and behaviors of life. Life manifests consciousness, and consciousness protects life within cellularities. Life is aware of itself and organizes itself to maintain, reproduce, and evolve life—all to protect and propagate an evolving code.

Just as the quantum world may not be strictly deterministic and may allow for variability in the form of multiple possibilities, the conscious world may not be strictly deterministic and may allow variability in the form of multiple possibilities and conscious choice.

Notably, when the dance stops, when all inorganic matter and radiation is compressed back into a singularity, then space and time collapse, but gravity persists. Which begs the question: when the organic—life—collapses and dies, does consciousness persist?

Governed by the laws of physics, the collapse of space-time into a singularity suggests that a singularity rests as a seed ready to germinate in another explosion and expansion of a new universe. Space, time, and gravity return to dance again, perhaps with life and consciousness once more, too.

Multiple singularities conceived from multiple blackholes, such as within this universe, suggest the probability of multiple universes. Therefore, at the very least determinism is more complex than knowing the coordinates of a single universe. Within the multi-verse, one would need to know infinite coordinates to determine finite outcomes within a single universe of a multi-verse. Yikes! More importantly and more likely, multiple universes suggest that variability and possibility is likely infinite, thus including randomness.

Contemplate this: In the multi-verse, universes would look less like air balloons expanding into relative nothingness and more like adjacent water balloons within a foam structure resembling cells in tissue. Conceptually, a solitary universe inflates, then deflates. In a multi-verse the universes have squishy interfaces. Some may be young and growing and some may be old and shrinking. One universe may indent, invaginate, or even leave vesicles within another, thus having the effects of matter and energy on its neighbor, but from beyond the veil and operating in darkness. Yet, this would cause deformation and displacement within the cosmoplasm of one universe, reciprocally affecting other universes. This could account for the inconsistencies in expansion across this universe and possibly explain dark matter and dark energy—hmmm.

One universe is sculpted by another and vice versa, and therefore one universe effects the next and the next, one is connected to the whole. Infuse the dimension of life into these worlds and perhaps a universe looks a little less like a water balloon and more like an amoeba where alteration in form and function is less passive reaction and more intentional creation—hmmm, again. A solitary beautiful balloon may not fit this dimension at all.

Is there a parallel to our present sense of genomic determinism that suggest that the genome determines life? This is a genome-centric view of life that is surely flawed. We know that the genome is nearly inorganic without the environment to dance with it. We also know that life is dependent on other naturally occurring phenomenon beyond the genome—and perhaps the environment is different in form and force without life to engage and observe it. This raises the question of whether we have a universe centric view of our world that is also flawed. Have we failed to include the possibility that the universe has interfaces at its boundaries that influence its internal behaviors and its form and force? Again, could it be that the universe is more like a cell within an infinite tissue than the infinite itself? It seems human centric and myopic, with a dash of arrogance and hubris, to think otherwise.

Life infuses creation, order, and purpose into the world of reaction, disorder, and inconsequence. The inorganic is made into the organic with an intent to protect and propagate the code of life. With this intention life organizes the energies of the cosmos within a metabolism for its purposes. Earthly life has purpose that seems beyond the stars and galaxies, even as glorious as they are.

The inorganic world is active and reactive but holds an equilibrium of energy where total energy is preserved, neither created nor destroyed. The inorganic universe tends towards entropy and disorder. The organic—life—reverses entropy and moves towards order, a push against the energetic equilibrium within the universe. Life's defiance of the traditional laws of physics and the concept of conservation, and equilibrium of energy manifests itself in a number of ways: for example, in the likes of flocks of birds flying in formations and schools of sperm flagellate swimming towards an egg. In fact, if life moves into this inorganic equilibrium, it undergoes dissolution and dies. A priority for life is to stay in disequilibrium relative to the inorganic world, further complicating the concept of homeostasis and a drive for internal equilibrium.

Within this biologic scenario one must also consider whether organic functions are just perturbators within the inorganic world. Or could life be much like a virus that infects a cell in that life is an "infection" of the universe? Could the molecular code for life be an endosymbiont to this universe? That would really be weird, but not implausible. The code of life being implanted within the universe, not evolved from the universe, would be a major disrupter of determinism.

Life also changes the universe. Life is aware of itself. Consciousness is necessary for the protection and propagation of the code. Through consciousness, life also observes the universe. Depending on life's vantage point, reality may look differently and there are infinite vantage points, thus infinite realities. But more significantly, the quantum world demonstrates that simply the act of observation changes the quantum world. Wave functions collapse into particles, energy becomes particulate, radiation becomes matter. Within this flow, life's genome and functions change the exposome. But by how much? The bidirectional balance is not clear.

Emotions, behaviors, and thoughts flow forward from physiology, and influence human social contracts, created ideations, cognitive constructs, symbolic narratives, and held beliefs. Yet, in a bidirectional fashion the social contracts, created ideations, cognitive constructs, symbolic narratives, and held beliefs directly affect emotions, behaviors, thoughts, and systemic physiology. These are inputs into the interosome that eventually feedback into the exterosome—inseparable from the whole. These advanced human neocortical functions that are relatively unique to the human species, at least on this planet, push back to influence the greater system.

Thus, human cortical awareness changes physiology. Perhaps even the neocortical awareness of being aware, the observer observing the observer, the sensor sensing the sensor, does the same.

As noted, fundamental in physics are the notions of multiple realities based on relative vantage points within space-time and even how the influence of simply observing an event within space-time changes the state of the observed. Life, and even life's observations, prove to be perturbators to all inorganic and organic systems, as well as the universe itself. The functions of the organic world influence the functions of the inorganic world.

"Out beyond the ideas of wrongdoing and rightdoing is a field. I'll meet you there."

Jalal al-Din Rumi

Contemplate: matter manifests the field of gravity to deform or warp the fabric of space-time. Life manifests the field of consciousness, while life's observing the fabric of space-time, sensing space-time, and then giving coherence to space-time, causes waveform collapse and particles to appear, energy coverts to matter. Within this context, life defines events and restricts time to present and a unidirectional future. Therefore, life manifests not just consciousness, but also matter and thus gravity through its observations. Thus, life contributes to the deformation of the space-time continuum. Within conscious, quantum, gravitational, electromagnetic, or even Higg's fields, what is primary—the field or what manifests the field? Chicken and egg? Connection and coexistence are, however, clear. Are these fields everywhere and all the time waiting to be manifest?

Notably, quantum waveforms have possibilities that speak to the future. Whereas the collapse of these waveforms into particles removes possibilities to create a moment in time—the present. Theoretically, at the speed of light time stops. Going faster than the speed of light arguably reverses time to create the past. Life cannot exist at these high speeds, but life and life's observations do seem to play a role in the creation of particles, timeframes, and the present. Once the present and the particle are defined time seems irreversible. Life creates the present, moves only towards the future, and seems to prevent moving into the past—perhaps a lesson on how to live.

The present exists because of observation. A point in time is defined by observation while energy appears as matter under observation. The world of reality experienced is therefore the relationship of space-time and life's observations, life's consciousness. The space-time world without this consciousness is different than the emergent life-space-time world experienced as moments manufacturing multiple realities to be remembered but not revisited. The constructed memory or imagination is neither the past nor the future but another moment in the life-space-time continuum—the reality of that moment but not the reality of the past nor the future.

Life is significant, yet one must concede that life is more of a perturbator and not a controller. A pinball player with small levers and careful nudges to influence outcomes—some more skilled than others.

The big concession to make is that human lives do not have very much control in the world, but humans do have some choice. Choice is most available to humans when not distressed, reactive, and impulsive, but on the contrary, when relaxed, engaged, and contemplative—when the neocortex is most fully online.

Threat feeds relative determinism to take away choice, and safety counters back with indeterminism, choice, and free will—yet without much overall control. Once a choice is made that moment in time has passed and possibilities for the future return, over which humans have little control—but choice comes again. Humans thus have the free will of choice, but very little control.

The philosophy of strict determinism (which seems on shaky ground—will leave religion for another space and time) concludes if individual human beings do not have free will, then they cannot be held responsible for their actions. If the idea that all was predetermined and one could predict all the following events from the point of universal inception for all time, including human behavior—then any accountability for human actions certainly seems absurd in concept.

But both the constructs of strict determinism within physics and philosophy seem intuitively incorrect to observed human behavior (although that should not be the standard; see relativity and multiple realities). Within the recent evolution of physics, strict determinism seems to have already been deconstructed. Yet it continues to rear its head within philosophy, religion, and mathematics...physics too (not dead, yet). Certainly, the addition of biology to the equation decomposes the probability of rigid determinism.

"Mathematics is construct, philosophy is narrative,
religion is belief, physics is metaphor, biology is reality."

Ali F. Oeus

Within a biologic model, determinism appears relative. Most, notably within biology, determinism is most relative to threat rather than safety. Threat and human threat physiology make humans more reactive, impulsive, obsessive, ritualistic, rigid, or stuck, and deterministic. Under chronic threat, adaptive patterns and habits can escalate into uncontrollable rituals and beliefs, even addictions. By contrast, safety and human safety physiology offers contemplation, curiosity, creativity, flexibility, planning, judgement, and time for multiple possibilities from which to choose. In other words, free will.

Awareness, action, consequence, and adaptation are some biological and fundamental functions of life that create changes from the neocortex to the genome to the cosmos. Accountability, an act of compassion and love, provides boundaries that keep humans safe, thus restoring the ability to choose wisely. In this way, holding each other to account, but in a supportive and empathetic way, also pushes back against determinism. The more punishment and threat, the less free will. The more accountability and safety, the more compassion and love, the more free will.

Have understanding and empathy in knowing that when humans are in threat they are inflicted by determinism. They may not be able to make a good choice or control what they do. And when someone makes a hurtful or harmful decision and action, accountability is a compassionate and loving act to restore safety.

With regards to determinism, conclusions include:

- The world may tend to be deterministic, but it is not strictly governed by determinism.
- Determinism appears only relative and not absolute.
- Threat increases determinism and decreases free will to choose.
- Safety decreases determinism and increases free will to choose.

- Accountability restores safety for all, thus increasing free will to choose.
- Therefore, determinism does not excuse accountability, but demands accountability.
- Choice should not be confused with control.
- The particulate self is manifested by threat.
- Radiant connection is manifested by safety.
- Just as life's observations alter physical energy forms from radiant confluent waves to distinct isolated particles, life's threats alter life energy forms from radiant confluent connection to distinct isolated selves.

Will biology's consciousness, observations, order, purpose, and choices be the organicity that dirties mathematics' and physics' sterile, inconsequential, yet calculatable universe to finally destroy the construct of strict, fundamental, determinism?

As life manifests consciousness, possibility, and relativity, could the force of life be the missing variable in the equation, the wobble in the space-time continuum, that connects the micro quantum world and the macro relative world?

TBD

"A human being is a part of the whole called by us universe, a part limited in time and space. [We experience ourselves, our] thoughts and feelings as something separated from the rest, a kind of optical delusion of our consciousness. This delusion is a kind of prison for us, restricting us to our personal desires and to affection for a few persons nearest to us. Our task must be to free ourselves from this prison by widening our circle of compassion to embrace all living creatures and the whole of nature in its beauty."

Albert Einstein

Threats

Threat is complex. Especially for humans. In humans, threats can come from both the interosome and exterosome. Threat can be acute or chronic. Threat can be cumulative and compounded. Threat can be historical, predicted, potential, real, perceived, remembered, or even imagined.

Review of Threat

The following is a long but not exhaustive list of threats. There is some overlap within the categories, but the goal is to be inclusive when completing a review of threat assessment. It is felt that the most of the 100% of clinical and nonclinical determinants of health and wellness can be found within this list.

Generational Threat – Genetic and epigenetic coding.

Historical Threat – Genetic, epigenetic, neural coding, and memories.

Ecological Threat – Heat, cold, storms, floods, radiations, toxins, pollutants, antigens, allergens, and pathogens.

Physical Threat – Lack of good nutrition, lack of shelter, drugs, alcohol, competitors, predators, abusers, accidents, and injuries.

Emotional Threat – Hate, shame, guilt, embarrassment, envy, jealousy, loneliness, and anhedonia.

Behavioral Threat – Impulses, compulsions, distractions, and addictions.

Social Threat – Isolation, crowding, neglect, abuse, teasing, mocking, ridiculing, bullying, rejection, cancellation, and poor boundaries.

Mental Threat – Cognitive distortions, bad contracts, faulty constructs, false narratives, and rigid beliefs.

Cultural Threat – Misaligned and maladaptive norms for emotions, behaviors, appearances, thoughts, narratives, and beliefs.

Societal Threat – Disenfranchisement, discrimination, indignity, inequity, and injustice.

Institutional Threat – Threat incorporated into policy, procedure, and law.

Financial Threat – Poverty and debt.

Technological Threat – Weapons, media, and artificial intelligence.

Spiritual Threat – Self focus, disconnection from the natural world, and lack of a greater connection.

Existential Threat – Angst and terror.

Humans may be the only species on earth that can contemplate their own death and annihilation—a terrifying possibility.

"This is terror: to have emerged from nothing, to have a name, consciousness of self, deep inner feelings, an inner yearning for life and self-expression—and with all this yet to die."

Ernest Becker

This load is heavy and the distance long. An algorithm is needed that treats this entire TL to reduce chronic disease, disability, pain, and suffering.

"The journey of a thousand miles begins with one step."

Lao Tzu

Coded Threat

Threat is complicated and abstract, and as complex as the universe itself in some respects. Generational trauma, adverse childhood events, and other past personal threat and trauma all contribute to an elevated TL and a code of threat sensitivity. [51] In addition, humans are genetically precoded and biased towards threat. This is priori core information, or core knowledge, that is embedded within the cells by the genetic code. It is this code that defines the real purpose and subsequent meaning of life—to protect and propagate the code.

Throughout a lifetime, humans modify and add codes, from the level of the mitochondria to the cells to the tissues and to the organs of the body. This is posteriori core information and knowledge that is housed in all the cells throughout the body, not to be confused with the limited amount of information and knowledge stored within a human's neocortical brain.

From inception to birth threat codes and cellular information, awareness, knowledge, and intention are present, even if cortical information, awareness, knowledge, and intention are not yet present. It is better to be primed to suspect a prowling and preying tiger and be wrong, than to suspect a warm and welcoming friend, only to be surprised by a tiger. This is an important consideration, as threat is cumulative and compounds when it is without being countered.

In addition, some of the most overlooked sources of threat are from the interosome, not the exterosome. In general, the more threats experienced in life, the more coded, or biased, towards a TR a human becomes.

These functions are not unique to humans. Even species

without the complex neural networks and the advanced cognitive capabilities of humans must engage in threat surveillance. Every species is constantly surveying their environment for threat versus safety. This is basic to survival.

As it is not possible to go through life without experiencing threat, this can become problematic. Physiology can be more easily triggered into an amplified TR based on cumulative and compounded past experiences, particularly if a TR is not actively decoded and a SR actively recoded. Organisms become sensitized, sensitive, predictive, and more reactive to coded threats. This has major implications for not only illness and disease versus wellness and health, but also for aging.

Organisms use past experiences and coding to predict future threats. Humans are wired in past threats to avoid future threats. The human aware cognitive mind is just a small part of this process. This speaks to why being present and being mindful is so challenging when humans are in threat. It also explains why cognitive behavioral therapies are only marginally effective. When humans are in threat, it is counter to physiology to be cognitive and fully present. Fighting biological drives creates conflict and threat physiology of its own. Therefore, the path to healing is primarily through decoding threat and secondarily recoding safety, moving from threat to safety, not through the control of the mind or modification of behavior.

Also consider, that those who have been coded or biased towards threat more than others become more sensitive to threat. Within primitive and tribal life, these folks served as sentinels for the tribe, warning that something might be amiss—attend and be vigilant, be aware or be lunch. A canary in a coal mine so to speak. The world is full of threat manifesting as anxiety, depression, disease, disability, pain, and suffering. All on the rise. These sentinel canaries are being ignored.

"I am almost convinced (quite contrary to opinion I started with) that species are not (it is like confessing a murder) immutable."

Charles Darwin

It is worth taking a moment to discuss what a code is. Information comes into the code. The code in response to the information, the input, produces an output which could be appropriately call outformation. For example, the exposome gives information to the genome that then creates outformation, phenome. DNA is the primary code that creates RNA as a secondary code that then is used to assemble proteins that become signalers, enzymes, structures, etc.—outformations. The code is the intermediate receptor and processor between input and output.

The genetic code is relatively stable, but not completely stable. Variation in output comes from the variation in input or from the changing of the code. Bad output can come from bad input or bad code. Misinformation and disinformation are forms of malinformation. Malinformation and mutation can lead to maloutformation or malformation.

This is not specific to humans, but to all systems. The intention of the human genome is to protect and propagate the code. Other systems may have codes with very different intentions that can conflict with the human genetic code. For example, other species' codes are to protect and propagate their code, not human code. A real tragedy is when humans build codes for other systems that do not support the primary goal of protecting and propagating the human species. These codes can be designed to protect and propagate a contract, ideation, construct, narrative, or belief that can compete or conflict with human needs.

When this happens, deconstruction and reconstruction,

decoding and recoding, needs to happen. But first, humans must be aware of this building of codes all the time and everywhere within society, culture, institutions, and minds that compete and conflict with the needs of the human species.

"Of course, bad code can be cleaned up.
But it is very expensive."

Robert C. Martin

Genetic Coding

Humans do have some threat instincts. Fear of heights—maybe snakes and spiders? Humans also have some reflexes that respond to threats—noxious smells, bright lights, loud sounds, disgusting tastes, excessive pressure, extreme heat, penetrating sharpness, etc. Additionally, the genome itself may be less fixed and more plastic than imagined and may undergo subtle changes in response to threat versus safety. This is most notably reflected in changes in selective nucleotides within genes, transposable elements or "jumping genes" that move location to create new code and change the transcription of the code, along with the changes in the telomeres capping the ends of DNA strands that shorten in threat and lengthen in safety. It is estimated that around 50% of human genes are transposable—so much for genomic stasis and a static real-self.

Most recently, it has been identified that in threat states mitochondria export their DNA into the cytoplasm of the cell to then be taken up by the nucleus of the cell and subsequently incorporated into the genome, not unlike a retrovirus incorporating its DNA into the genome. The genome is both plastic from an internal but also from an external perspective all along the threat-safety spectrum

Genomic plasticity has been demonstrated in a variety of diseases. Short lengths of DNA repeated multiple times, "tandem repeat expansions," are known to correlate with more than fifty lethal human diseases, including amyotrophic lateral sclerosis, Huntington's disease, and multiple cancers. Thirty per cent of these diseases have at least two common alternative forms of a gene caused by these mutated repeated sequences that are located in the same place on a given chromosome. This does not appear to be random, but programmed mutation, that TVST suspects is a genomic response to chronic threats. The genome is dynamic and more than a template to produce proteins. It is an adaptable and intelligent program that modifies the code of life in attempts to sustain the code of life.

Upon exposure to environmental changes, genetic code appears to be able to write new code to allow for infinite possibilities in presentation.[52,53] However, by and large, the genetic code is relatively slow to change. Traits can change faster than genes change. It appears that most modifications of the code's expression come from extra-genomic modifications such as DNA methylation, histone modulations (proteins that interface with DNA to change expression. Different histone "tags" can even modulate further), and non-coding RNA adjustments, instead of direct genetic modification through nucleotide substitution and gene changes, i.e., most modifications of genetic expression are add-ons to the genetic code—epigenetic coding.

The code is not completely fixed and predetermined. It reflects the environment, dynamic and every changing in structure and expression. Genomic determinism is only relative.

Epigenetic Coding

Much of the human TR is learned, but not all in the classic sense of learning with informational processing at the cortex of the brain, subsequent mental constructs, memory consolidation, and then the retrieval of memories.

Instead, much of this learning is at a mitochondrial and cellular level, where both the mitochondria and cells go through genetic changes and genomic expression alterations via epigenetic changes. Epi means on or above—indicating these changes in expression come on or above the genetic code itself. They do not change the code, just the expression of the code. Therefore, epigenetic changes should be preventable and reversible.

The DNA methylation modifications, various histone modulations, and non-coding RNA adjustments that influence the moment-to-moment expression of the genetic code also create a "memory" of threat and a bias towards a TR based on past exposures and experiences with threats.[54] This selective expression of the genome is based on the exposure, the sensing, and then the signaling processes from both the interosome and exterosome—the exposome. The exposome interfaces with the genome, determines the epigenome, and thus, the phenome.

Epigenetic coding is influenced by human lifestyle and environmental factors. Epigenetic changes are dynamic and reversible but may be robust if the environment is without change. Epigenetic changes can be inherited, as well. These changes affect the expressions of inherited genes without changing their sequence.

Epigenetic changes can "turn on" or "turn off" a gene through various mechanisms, including DNA methylation modifications, histone modulations, and non-coding RNA adjustments. Generally, methylation blocks gene expression, and demethylation unblocks to allow gene expression. Histones are chromosomal proteins wrapped by DNA. A tight packaging of histones prevents the access of transcription factors to DNA, inhibiting gene expression. A loose packaging of histones does the opposite. Histone tagging refers to adding or removing chemical groups from histones to regulate its tightness or looseness of packaging.

Rat models have demonstrated that pharmacologic interventions which target transcriptional mechanisms can rescue DNA methylation modifications associated with early life distress, improve threat-related trajectories, and even decrease risk for cancers with aging. Environmental threat versus safety plays a major role in these coding processes.

Both the genetic and epigenetic coding with biases towards a TR can be passed forward for generations. Generational transfer of epigenetic code makes sense. If the world is hostile and threatening, it is advantageous to come to the world precoded and biased towards protective phenotypes and physiologies. However, as these protective physiologies in chronicity have increasing toxicities, they have significant implications with regards to illness and disease versus wellness and health over a lifetime, especially in populations who have suffered through many years and many generations of sustained threat.[55] Threat coding and threat disorders, including mental illness, whether from maternal or paternal codes can adversely affect offspring development even within the womb.[56]

Neural Predictive Coding

In addition, the neural structures below the level of the human brain cortex code the experiences of threat. Threat thus biases neural wiring and creates predictive codes oriented towards a TR.[57] Here, cellular genomic and epigenomic functions are at play within the cells of the brain, most specifically the neurons and the glial cells.

As humans evolved from less sophisticated life forms (archaea with endosymbiotic bacterial and viral descendants and remnants) that performed surveillance via cellular awareness without neural structures and cortical brain awareness, it makes sense that humans also conduct much of this surveillance below cortical brain awareness. This occurs not only in the wiring of subcortical neural structures, but all the way down to the neural cellular and mitochondrial chromatin (DNA-RNA complexes with all their attachments and modifications).[58]

Memories and Dreams

Memories are another form of coding that can intermittently bubble up or be called up into cortical brain awareness. They are different than predictive codes, as they come to cortical brain awareness—sometimes a foggy awareness but nevertheless cortical awareness. They, too, incorporate epigenetic and predictive codes in their manifestation. The epigenetic coding of memories means at a certain level they may be transmissible across generations.

Memories are almost immediately influenced and modified by previous codes. The reality in the moment of memory formation may not be the same as the reality within short or long-term memory consolidation as these memories have already been modified by the state of the system.[59] Unpleasant or traumatic memories can activate a TR.[60]

Cortical processes that re-enforce memories, such as rehearsal, tend to be competitive and extinguish other memories, but subcortical processes that re-enforce memories and connections appear collaborative in that they improve other memories and their associations. This later type of memory is very active while humans sleep. This has implications with regards to study habits, but perhaps more importantly are the implications within psychotherapy and trauma therapy.

Dreams are pulled forward from previous codes and memories. Unpleasant, traumatic dreams and nightmares can arise from a platform with a high TL and activate and accelerate a TR.[61]

Memories—correct or incorrect—and dreams, as with predictive neural codes, are housed within genetic coding, mitochondria, cells, neural networks, and the brain. Within the medial temporal lobes of the brain, hippocampal neurons demonstrate DNA breaks, nuclear release of DNA fragments, and inflammatory changes for hours after learning tasks. Humans wear down and exhibit symptoms of a TR with prolonged cognitive loading. This requires periods of recovery to resolve inflammation, repair structures, and to feel well. Memory consolidation is best served by periods without neocortical activation, out of threat, and within safety—sleep being a priority state.

All biological code eventually leads to the genetic code.

Occult Threat

Perhaps the most overlooked sources of human conflict and threat physiology come internally from emotional, behavioral, and thought constraint, repression, and suppression.

The stuffing of aversive emotions, restraining of behavioral drives, and compartmentalization of negative thoughts—particularly if these emotions, behaviors and thoughts are never allowed expression—register much like other conflict and threat within the human brain and activate a TR.[62] The neocortex of the brain and its contained contracts, ideations, constructs, narratives, and beliefs, when in opposition to physiological drives and desires, can create conflict.

Raw emotions are generated from the periaqueductal grey matter of the midbrain, located at the top of the brainstem. This is in close proximity to the brain nuclei that produce the neurotransmitters dopamine, noradrenaline, and serotonin. The signaling from subcortical structures, particularly the midbrain, moves upward and forward towards the cortex of the brain through the striatum to its tip—the nucleus accumbens—to drive emotions, behaviors, and thoughts. The neocortex through the medial prefrontal cortex can temper these responses, or inhibit this flow of physiology, thus blocking the feeling and expression of emotions. In this way, raw emotions can be managed or repressed, yet the physiology isn't completely extinguished. It can even be amplified at levels below the neocortex. This process creates conflict between the formations and needs of the human neocortex and the drives and desires of the more primitive physiology bubbling up from below.

This conflict activates a TR throughout the system from the cortex, particularly at the anterior cingulate cortex, to the subcortical limbic structures, to the brain stem, and eventually all the way down to the cells. In fact, when this physiology is not allowed neocortical sensing, expression, and integration, the cells may be more aware of the conflict than the neocortical brain.

When threat is severe, chronic dissolution can occur at the mid brain, with neurotransmitter depletions, numbing loss of emotions, loss of motivation, motion, and movement, social disconnection, an altered sense of self, cognitive fogginess, and nonconsciousness. Pyschostimulants, risk taking, and self-mutilation (such as cutting) can still induce a neurotransmitter response, even within a depleted system that can bring a sense of "normalcy" back, temporarily. Recognize the system is doing exactly what it is designed to do, given the predicted, potential, real, perceived, remembered, or imagined threats. Safety resolves all this without risk.

Traditionally, gaba-amino-butyric-acid (GABA) has been associated with relaxation or even sedation. But the reality of GABA is that it is an inhibitory neurotransmitter (inhibiting or hindering neurons) and that it is contextual to threat versus safety. In conflict or threat, GABA has inhibitory roles over the striatal emotional, behavioral, and thought flow, as well as the functions of the neocortex. In this scenario increased GABA levels in certain brain areas and this connectotype are associated with increased distress, emotional, behavioral, and thought constraint, repression, and suppression, and anxiety, particularly social anxiety. Neurotransmitters always need to be looked at within the context of threat versus safety to clearly understand their effects. It is worth remembering that cells knew neurotransmitters such as dopamine and noradrenaline well prior to the invention of brains.

The brain is recoded and restructured significantly during

adolescence. Explicit learning, especially around social survival, is associated with this neuroplastic period of development, and so are the development of mental health issues. There is some acute social benefit to constraining, repressing, and suppressing emotions, behaviors, and thoughts, but in chronicity there is also some potential toxicity that make adolescence a time of vulnerability that requires great care.

Anger, fear, and sadness are valuable emotions—physiologic states—that signal the need to change, redirect, and move in a different direction. This is not dissimilar to a primitive bacteria or archaea sensing an acid flow within the primordial soup, then moving to a safer place. These states need to be felt, acknowledged, integrated, and used to correct course. They should not be inhibited, denied, constrained, repressed, suppressed, or wallowed in, any more than ancient life forms should stay within an acid flow.

There are many social and cultural reasons and potential benefits for acutely repressing, constraining, suppressing emotions, behaviors, and thoughts. But there can also be tremendous costs in terms of illness and disease to what chronically goes unfelt, unacknowledged, and unexpressed.[63]

"What we resist, persists."

Carl Jung

Codes of Human Construct

This section would not be complete without commenting on codes of human construct that can support threat physiology. Misaligned and maladaptive codes of conduct, policy, procedure, and law can become culturally engrained and institutionalized, resulting in great harm.

Similarly, code can be created within computer algorithms, such as for marketing, social media, and artificial intelligence.

These non-biological codes can be misaligned, maladaptive, and manipulative—"dark patterns." They prey on subcortical, subconscious human neurobiology, threat coding, threat biases, and threat behavior to harm an individual for the power, prestige, and profit of the perpetrator. The insidious to occult nature of these type of codes of threat help the perpetrators to survive, if not thrive. This modern form of coding is a form of a virus infecting and weakening host individuals and societies for its own advantage, propagation, and perpetuation.

These artificial technological codes in many ways seem to be the antithesis of the natural spiritual codes of the species. Modern technologies are pulling humans further away from evolutionary wiring and the therapeutic connections of the past. These technologies desperately need regulation, and humans need to build an immunologic network that can defend against these advancing technologic threats. The infusion of artificial technology into broadcast and social media make them highly weaponized and risky to human beings. The future under the influence of AI remains uncertain, but in many ways, humans seem to be perched on the edge of their own destruction.

All new technologies, whether 20th century nuclear, 21st century artificial intelligence, or 250,000 year old "sapiocotical" evolution—whether inorganic or organic technologies—have the potential for bad or good, destruction or creation, and threat versus safety. Both the human genome's ability to write its own code when under threat and artificial intelligence's ability to write its own code to create threat and unplanned outcomes are of concern. There is a certain irony in the current concern that the human creation of artificial intelligence and its ability to write its own computer code and propagate itself may lead to the destruction of the world as we know it today. Meanwhile human intelligence and the human genetic code intended for the perpetual propagation of the species are already well on their way to inflicting this destruction through chronic threat and the associated disease, disability, pain, suffering, and death. This seems particularly twisted and certainly elevates the level of threat within the world. Humans probably need to worry more about human behaviors than computer behaviors at this point, but it is fair to say that the "sapiocortex" and AI can each cut both ways for good or bad. Both need safety-based coding, not threat-based coding.

"Success in creating AI would be the biggest event in history. Unfortunately, it might also be the last.... unless we learn how to avoid the risks."

Stephen Hawking

Recognize that AI is not the first code humans have built that codes itself. Institutions are built on information interfacing with codes of human ideations, contracts, constructs, narratives, and beliefs. These many institutions include education, science, environmental care, criminal care, health care, public health, global health, social welfare, law and justice, economics, business management, multi-media, foreign affairs, and defense, to name a few. But only some of these institutions have the potential to be powered to write their own code to perpetuate themselves. Within business management, the concepts of capitalism and corporations deserve some reflection.

Inherent within the current code for corporations is the idea that the benefit of financial investment and growth is so high it warrants disregarding the risk of marginalizing total cost accounting. For example, the adopted ideations, contracts, constructs, narratives, code of operations and conduct decline to hold institutions and associated individuals to account.

The costs of ecological, physical, emotional, behavioral, social, mental, and spiritual degradation must be included in our accounting formulas to balance our books. Well-being, protection, and propagation of humans need to be prioritized over greed, production, and profits—a false profit so to speak. Democracy, transparency, fairness, well-being, and the restoration of biocapacity should be the guiding principles to lead us forward.

Corporations are not held to being responsible for their pollution and are given free passes on paying even their basic taxes. Exxon is not Exxon without the freedom to damage the planet essentially for free. Amazon is not Amazon without the roadways and airways provided essentially for free. Private equity is allowed to generate returns in the shadows without liability for poor outcomes. Social media is gaining great profits at the expense of others and the human species itself. Accountability is a necessity within TVST; there are no free lunches. The constructed codes have not held corporations to full account, and in many ways, have facilitated the growth of bad code.

Remember that the purpose of every species is to protect and propagate the genetic code of their species—the composite of individuals and their progeny—the group. The United States Constitution aligns with and protects this purpose. The false-self does not. The goals of the false-self are to acquire and accumulate prestige, power, and profit for the individual. Neither does the corporation, another form of false-self, align with the purpose of the human species. The primary goal of the corporation is profit. The corporation's status as an individual has been errantly codified into law by the United States Supreme Court. This gives corporations undue extended opportunities for prestige, power, profit, and influence.

How the code is written is how the corporation responds to information provided to best achieve this prestige, power, and profit. Input to output, no evil, just process. Yet, the more prestige, power, and profit a corporation has, the more influence it has, and thus the more it can lobby to write its own code to perpetuate its prestige, power, and profits—this is dangerous.

The United States Supreme Court has given corporations the rights of the individual, "personhood," without any clear biological or constitutional basis for this determination, and without the necessary nuance and guardrails. In other words, the codes are lacking to adequately restrain and hold corporations to account. Congress, or better said, congress people, have acquiesced to this determination. Bad code programs worse code.

It would have been better if the SCOTUS had given corporations the purpose of human life, the purpose of the human species—to protect and propagate the human genetic code—instead of a false "personhood" and the rights of the individual human. Written well in advance of the Bill of Rights and this erroneous SCOTUS decision, the Preamble to the Constitution outlines this priority to the species.

The SCOTUS has essentially created a new nonbiological species much like AI. Unfortunately, this species has evolved to be both a competitor and a predator to the human species. Like a virus inflicts the damage it does without malintent, a sense of fault, or any guilt, the corporation does what its code allows or dictates without malintent, a sense of fault, or any guilt. It's just input to output. However, unlike a virus sustaining its code, humans write the corporate code. Given that humans write this code, it is preferred that it be written to support the purpose of the human code, and to have corporations live symbiotically with the humans. Humans have control over this alignment. Corporations can be restrained and held to account—time will tell if AI will be restrainable or accountable to humans.

This is not to say corporations, or any institution for that matter, are necessarily evil. However, within the process of building and regulating these entities they should be viewed not as individuals, or preferential to individuals. Rather, they should be viewed *in service* to individuals, societies, and the species. Corporations should be transparent and held to total cost accounting along the spectrum of threat versus safety and not just accountable to profit versus loss.

Capitalism uses the motivation for profit to drive behavior, efficiencies, and improve resource allocation within a transparent and free market. Transactions within this market are thus felt to be mutually beneficial and fair. Total cost accounting must be used to set prices properly. Again, no free lunches. Government's only role in business affairs is to ensure the markets are transparent and fair. Alternatively, corporations' only role in government affairs and elections is...none. Allowing corporations access to the mechanics of governance is to allow them to write their own code. It thus becomes a code solely based on prestige, power, and profit, with no awareness of the need to protect and propagate the species.

Capitalism and corporations are not inherently the problems of modern civilization. The problem lies within the bad codes that have been written that fail to set proper boundaries and fail to hold to account. Capitalism allows for big ideas to flourish and precision and efficiency within markets. Corporations provide a service to individuals, societies, and the species by producing needed or desired goods within a hopefully transparent and free market. The mark or measure of success in this mission is a fair profit for all those who take risk and invest. This is potentially good for all humans. Corporations with excessive prestige, power, and profit that write their own code is bad for all humans.

Corporate AI is another story. AI is also a nonbiologic species created by humans. Will it be a cooperator and collaborator or a competitor and predator? Unlike non-AI corporations that are dependent on humans writing code for them, and thus their behavior can be modified, AI can write its own code. Once up and running, humans have less, and potentially nothing, to say in how AI aligns with the human species. AI's unregulated functional code gestating within the corporate womb and in relation to the corrupted corporate code has the potential to birth something exponentially horrific for humans and the planet. Threat is associated with preterm births, preeclampsia, and birth defects. Unfortunately, under competitive threat AI has been birthed prematurely with notable defects, distress, and hallucinations. What will AI become—symbiote or pathogen, friend or monster?

Notably, corporate codes applied to health care on total have been an abject failure in improving the health and welfare of individuals and societies. Will AI follow suit? Hopefully, in the future, corporate and AI code can be written that follows a similar path of biologic evolution where conflict and competition take a back seat to cooperation and adaptation to allow the human species to go beyond just surviving to assist in thriving.

Investment in AI offers opportunity for enormous corporate profits, whereas investment in people offers enormous societal benefits and cost savings. A sound social safety net will do more for the health and wellness of people and societies than AI will ever do.

"Earth provides enough to satisfy every man's needs, but not every man's greed."

Mahatma Gandhi

As humans are biased towards threat coding and threat physiology over safety coding and safety physiology, it is easy to accumulate threat codes. It is important to actively and continuously decode threat, and proactively and continuously recode safety on all fronts to maintain a physiology that sustains health and happiness. Most humans are unaware of their total TL, let alone how to decode threat and recode safety. Humans also appear to be relatively unaware of the mistargeted and bad code they create that further threatens the species.

Threat Load (TL)

Threat Load = Active Threat +
Coded Threat + Occult Threat.

The TL determines the sensitivity, severity, and chronicity of the TR.[64,65] A TR to an acute event such as an infection, insult, or injury determines acute pain, illness, and incapacitation. A TR from historical and sustained threats determines chronic disease, disability, pain, and suffering.

To fully understand a human's (and society's) risk for illness, disease, disability, pain, and suffering, the total TL needs to be understood. Few humans have this understanding, including health care providers.

Threat Sensing and Signaling (TS)

In the primordial soup TS is related to the exterosome. Primitive life forms became adept at sensing perturbations within the environment that were threatening—too much pressure, too hot, too cold, too acidic, too alkaline, toxic chemicals, toxic gases, other toxins, intense radiations, pathogens, and predators, predominantly.

In addition to sensing these perturbations in the environment, primitive life forms learned to sense danger based on cellular components within the primordial soup that wouldn't be present if conditions were safe. If an organism should perish, the contents of their cells, or the bits and pieces of their cells, can signal that the environment is unsafe. This, in turn, can stimulate a TR—colonize, attack, escape, hide, or become lifeless and dormant.

Abnormal extracellular concentrations of specific molecules—purines, pyrimidines, monoamines, polyamines, and peptides—can all signal threat. Also, free-floating molecules, composed of fat, protein, and sugar from the breakdown of cell walls, cell membranes, and other cell structures, can signal destruction, death, and danger, and activate a TR.

Eventually, more intentional, and direct methods of intra-species inter-organismal TS evolved using different modalities—radiations, electromagnetism, vibrations, purines, pyrimidines, monoamines, polyamines, and peptides. All these defense mechanisms served as precursors to how humans would eventually come to sense, signal, and respond to threat.

Humans, too, have receptors for light, shock, vibration, temperature, sharpness, pressure, chemicals, gases, toxins, pathogens, and predators to sense threat. Humans also use light, electromagnetism, vibrations, purines, pyrimidines, monoamines, polyamines, and peptides to sense and then signal threat. Human cells can sense what should be intracellular contents spilled into the extracellular space, along with the fragments of like and unlike destructed and decomposed cells (alarmins, pathogen associated molecular patterns—PAMPs, and danger associated molecular patterns—DAMPS[66]), as threat signals.

Admittedly, humans are more complex than simple single-cell organisms and their colonies, such as those that existed within the primordial soup. Nevertheless, humans are also still these simple cells (archaea with endosymbiotic bacterial and viral descendants and remnants) within the depths of human's genetically coded threat phenotypes and physiologies. Indeed, human embryos—multi-cellular, not yet differentiated-cellular forms—look suspiciously like bacterial colonies in many ways. However, human cells are more complicated than their primitive ancestors, as are human beings and human colonies. Yet, perceiving a human only as a uniquely siloed individual single entity, a disconnected singular whole organism, misses much of the evolutionary cellular and social complexity of what it is to be a human.

On closer inspection, humans are complex multi-organed,

differentiated-cellular, multi-cellular, multi-organelled, eukaryotic cell colonies. Human mitochondria, cells, tissues, and organs are highly specialized. Humans sexually reproduce as a coordinated system of cells (as opposed to autonomous asexual cellular reproduction of primitive life). Yet, humans are still just a collection of cells with individual cell functions that have deep evolutionary roots, including the functions of primordial prokaryotic single-cells and their colonies. The primitive phenotypes expressed in a spectrum of threat and safety physiologies persist in humans today.

What is different in human beings' TS is that humans carry the equivalent of the primordial soup within the spaces between the cells—the interstitium. Internally, this is where the soup meets the cells, the base level of sensing, signaling, and intercellular communication within the human interosome. It is critical in TVST to understand this internal intercellular environment.

Humans also interface with the exterosome—air, water, soil, food—which then influences the interosome and physiology. Epithelial cells of the skin interface with the exterosome. However, so do the epithelial cells of the respiratory, gastrointestinal, and genitourinary tracts. Other interfaces with the exterosome include: the nose, where sensory neurons are embedded within epithelial cells. The eyes, most notably the retinal cells, sensing light and color. The ear's tympanic membrane and small bones, sensing vibrations or sound waves. The mouth, with receptors for tastes and toxins. The skin's receptors for vibration, pressure, pain, and temperature. All these interfaces are the first layers of defense from external threats. A TR—combined metabolic, inflammatory, immune, and behavioral defense responses—can be initiated at the level of these cells well prior to the recognition of threat by the brain.

Humans have another protective interface with the environment that has importance—the vascular endothelium containing the blood. The endothelium of the vascular tree is a dynamic internal interface. It too is a potential barrier to the interstitial space, but also allows access to the interstitial space when executing a defense. These endothelial cells are very integrated into the TR. The endothelial cells provide a barrier to toxins and pathogens entering the bloodstream. They also control what exits the blood stream into the intersitium, tissues, and organs. In a defense response, both the epithelial cells and endothelial cells will loosen their junctions and connections to allow proteins and cells to migrate out to neutralize a threat.

Within the vascular endothelial walls, the blood itself is considered another protective connective tissue. To that end, it is packed with immune cells surveilling the environment and ready to be deployed in a defense. A threat and immune response, netting, trapping, and killing invading pathogens, can rapidly occur within the blood prior to a systemic invasion into the interstitial space and interstitial soup, and prior to interfacing and engaging all the other cells, tissues, and organs of the body.

The epithelial cells interface with the exterosome, and the endothelial cells interface with the interosome and the blood. Both serve as protective layers and barriers to the interstitium and the other cells, tissues, and organs. Both layers of tissue communicate back to the greater system the status of threat versus safety. Within chronic threat physiology, these protective cell layers of the skin, respiratory, gastrointestinal, genitourinary, and cardiovascular networks are represented in numerous noninfectious chronic diseases, blood vessel diseases, and autoimmune diseases.[67,68,69] Notably, 80% of cancers and 90% of cancer deaths are from cancerous forms of epithelial cells.[70]

It is at the level of the interstitium that most human cells are bathed. The interstitium is where much intercellular communication takes place. This perhaps best resembles the functions occurring within the primordial soup between ancient organisms. Understanding the interstitial environment and the intercellular communications within are essential to understanding human illness and disease versus wellness and health. In addition to understanding human evolutionary connections, understanding the differences between these cellular life forms and modern complex humans is also essential. Humans have eukaryotic cells, and complex multi-cellular, differentiated-cellular, and anatomic forms that distinguish them from simple single-cell life forms. Humans also differ from simple organisms in having complex emotional, behavioral, social, mental, and spiritual influences and needs.

The systemic TR to emotional, behavioral, social, mental, and spiritual toxins, predators, and trauma is interestingly very similar to a response to an allergen, antigen, pathogen, or physical trauma. Therefore, in understanding the complexities of human emotional, behavioral, social, mental, and spiritual functions, it is also important to understand the role of the cell and the environment in human illness and disease versus wellness and health.

Dissolution in Complex Life Forms

Another important concept to understand is the tendency for more evolved life forms to go through an evolutionary, or phylogenetic, regression when under threat to rely on more primitive phenotypes and functions in a defense state. This has been described as *dissolution*.[71,72] This phenomenon can be seen in both acute illness and chronic disease states as organs, tissues, and cells become more autonomous from the greater system (ex: the heart—arrhythmias, the brain—seizures, the epithelia—cancers).

In addition, complex organisms in a defensive survival mode use resource allocation programs that tend to downregulate and even deconstruct and destroy (autophagy and apoptosis[73]) less important structures to assist in a defense. These processes aid in and sacrifice for the demands of the fight, flight, or fever and the protection and propagation of the code. The shutting down of the gastrointestinal, genitourinary, and reproductive functions occurs when "stressed" or in threat. Additionally, there is an even more interesting phenomenon occurring within a sustained TR, where the more advanced "sapiocortical" social, imaginative, cognitive, language, and memory associated neurologic functions are undergoing dissolution, if not degeneration.

The human brain consumes up to 20-30% of the total energy produced by the system (varies with phenotype), while only accounting for 2% of body weight. The human neocortex consumes the majority of this energy within certain phenotypes. Parts of the phylogenetically newer functions of the human brain are downregulated and even deconstructed when in threat.[74] This dissolution of neocortical functions includes neuroceptive dissociation. It is most profound in the areas responsible for complex human mental and executive functions, declarative and factual memory, receptive and expressive language, and connection, bonding, empathy, and other advanced social functions. As threat escalates or becomes more chronic, the older cortical parts of human brain functions can also undergo dissolution including interoceptive dissociation. This results in somatic, visceral, and emotional disintegration characterized sometimes as disembodiment or simply not feeling like oneself. This progressive process culminates in further impaired functions and worrisome alterations in the sense of self with loss of energy and affect, and complaints of feeling distant and numb.[75]

Initially, with TS, the more autonomous subcortical structures are prioritized, such as the limbic and striatal networks. As TS escalates or persists chronically, even the immediate subcortical neural structures can undergo dissolution.

Dissolution of limbic, striatal, and even lower brainstem autonomic functions can be seen in severe and chronic threat states as central neurologic and regulatory functions give way to more local autonomous functions. The central control is gradually turned over to peripheral reflexive mechanisms under threat. Eventually, in extreme threat, regulation and function can be turned over to the cells themselves. This autonomy can lead to infidelity to the greater system.

The process of threat-associated dissolution has been characterized as evolutionary regression in some or dissociation in other models of illness and disease. Although not well-delineated to date, this process of dissolution is a major component of models for migraines, developmental disorders, neurodegenerative diseases, autoimmune disorders, cancer, and emotional, behavioral, social, and mental illnesses and diseases.[76,77]

Illness and Disease

Acute threat underpins acute illness, incapacitation, and pain. Chronic threat underpins chronic disease, disability, pain, and suffering. These concepts include not just physical illness and disease, but all emotional, behavioral, social, mental, cultural, and spiritual illnesses, diseases, disorders, and syndromes, including addictions.[78]

Progressive elevations in threat, TL, and TS initiate threat-associated epigenetic codes, phenotypes, and associated physiologies. Escalating, compounding, or chronicity of TS will move organisms from T1 to T2 phenotypes and their associated physiologies. This progression can be characterized as moving from combustion to conservation, or from high-energy demand states to low-energy demand states.

Under the influence of threat and TS, humans initially become progressively more catabolic, inflamed, irritable, reactive, selfish, antisocial-asocial, disconnected, and dissociated. The cellular immune system is activated within these early defense states. In chronicity, there is susceptibility to inflammatory, immune, and metabolic diseases. In addition, there are a plethora of associated emotional, behavioral, social, and mental impairments and disorders.

With an even progressively higher TL, chronic threat, prolonged exposure to threat, being trapped in threat, and increased TS, humans become further dissociated, catabolic, self-focused, isolated, disconnected, asexual, depressed, desperate, hopeless, and helpless. They can even progress into a state of physical and emotional collapse.

Chronically elevated threat, TL, and TS result in systemic mitochondrial, cellular, tissue and organ dysfunction and degeneration, autonomic dysregulation, chronic fatigue, depression, hopelessness, helplessness, with submission, subordination, surrender, sickness behaviors, and progressive dissociation.[79] At this far end of the threat spectrum, there is poor cellular immunity and an increased risk for more severe and chronic diseases. These include chronic infections, auto-immuno-inflammo-metabolic diseases, and the initiation, propagation, and metastasis of cancers.[80,81,82,83]

Threat-associated emotional, behavioral, social, and mental impairment, illness, and disease correlate and persist across the spectrum of physical illness and disease. However, these impairments are phenotypically different within these advanced immobilizing T2 states, as opposed to the earlier mobilizing T1 states. However, these physical, emotional, behavioral, social, mental, and spiritual changes and afflictions are all from the same root—threat—and signaled through the ever-changing physiologic soup.

These concepts lead to questioning the construct that "what doesn't kill you makes you stronger." Cumulative, compounded, and chronic threat appear to make humans weaker. Safety makes humans stronger and more resilient. Biologically, resilience is dependent on periods of safety, not repeat exposures to threat.

In addition, it needs to be considered in illness and disease models that those who have premature degenerative diseases and hyperresponsive immune and inflammatory networks may be those who mobilize internal resources to fight off pathogens and predators faster and better. Today, a genetic profile with a double APOE4 allele (normal variant of a gene) is worrisome, and used as a prognosticator, for neurodegenerative diseases.[84] APOE4 is associated with the mobilizing of resources, especially lipids, from neurologic structures. This gene may have had great survival benefits hundreds of thousands of years ago in the wild. Back then, mobilizing resources for a defense from a tiger, including fuel production and immune activity, was a priority. Therefore, this gene persists today.[85] Unfortunately, in the modern world, threat exists less commonly in an acute physical form and more commonly in a cumulative and compounding chronic multifactorial form. This once genetic asset appears to have become a liability to those in chronic emotional, behavioral, social, mental, cultural, and spiritual threat. This manifests itself most notably in those with degenerative neurologic diseases, such as dementia and schizophrenia.

Peptide Signaling

Peptide signaling is one mechanism of intercellular communication that can be traced to better understand TVST.

The inception of the peptide communication network dates back to primitive archaea and bacteria—the first life forms on Earth over 4 billion years ago—that used intercellular peptide, "autoinducer"[86], signaling for basic communication with other cells within the primordial soup.

Peptides are short chains of amino acids. They are essentially small proteins. Peptides are produced by transcription and translation of the genetic code for the numerous essential functions of an organism. Peptides can be incorporated into cellular structures, used as enzymes in cellular metabolism and defenses, and as signalers to other cells, amongst other things. Peptides can be formed in outer space on the condensation of cosmic dust and are likely to be as old as the universe itself, estimates range from just over 14 to as much as 27 billion years old.[87] In fact, the environment of outer space appears more conducive to the spontaneous formation of peptides than the environment of Earth.

It is possible the purine/pyrimidine network arrived from outer space in short chains of RNA as well. The purine/pyrimidine building blocks for RNA and DNA have been found in cosmic debris. In addition, chains of nucleic acids have a propensity to recreate spontaneously. Thus, RNA may have hijacked primitive cell metabolism and structure for its own reproduction, akin to modern viral RNA reproduction. In time, with evolution, these nucleic acid structures may have adapted to reside permanently within a cell as an endosymbiont in a more stable form, DNA.[88]

Arguably, peptide signaling may be slightly more modern than purine and pyrimidine signaling, such as the signaling of extracellular adenosine triphosphate. (Biologically, eATP (a DAMP) looks suspiciously to be the primary signaler of threat.) Therefore, perhaps peptides are secondary to purine and pyrimidine primary signaling of threat. Alternatively, peptide and purine/pyrimidine signaling may have evolved simultaneously within the primordial soup. Amino acids and peptides are easier to form, tend to self-aggregate, sometimes self-replicate, and are more stable than nucleic acids and their structures. Peptides seem to provide stability to nucleic acid molecules and their structures as well.[89] Notably, under conditions like the primordial soup, self-assembling peptides can aggregate into fibers, known as amyloids. In an unknown way, they direct order to RNA and DNA assemblies. This all is a prebiotic process that gets very close to the mechanisms for life on Earth.

Although the construct for the evolution of life has leaned heavily on the concept of competition and the survival of the fittest, repeatedly cooperation and adaptation are seen to be more significant players in these processes. This seems to be the case even in the prebiotic activities of peptides and nucleotides. Coevolution and interdependence may be the proper answer to the "chicken and egg" question of whether peptide or nucleotide signaling is primary versus secondary.

Regardless of what came first or second and from where, the peptide story is easier, clearer, and purer to tell and follow than the purine/pyrimidine story when explaining TVST. Therefore, the peptide narrative has been used over the many other possible communication networks, including the purine and pyrimidine network and the bioelectric network, as well as the phylogenetically much newer neurologic, autonomic, endocrinologic, immunologic, and complement networks, and the choline, monoamine, polyamine, other peptide, and steroid ligands. However, none of the networks and ligands are separate from each other. They all work together to support the system. To accept the peptide story as THE story in the complex systemic physiologic flow of a human in threat versus safety is too simplistic. Suffice it to say that the entire system is coordinated in its response. The peptide network is just a representation of the system, not THE threat and safety signaling system itself.

Specifically, the cytokine-peptide network will be used as a tracer within this work. Again, this is not to dismiss all these other networks. These other networks and their ligands, as well as noncytokine peptides and the cytokine network, are all integrated in a TR or SR, frequently in bidirectional signaling. [90,91,92,93]

The Cytokine Network

The cytokines are newer (~2 billion years old), larger, and more complex signaling peptides than the 4-billion-year-old autoinducers of primordial single-cellular life forms. The cytokine network evolved with more complex multi-cellular life forms a little more than 2 billion years ago. Examining the cytokine network will help with understanding threat sensing and signaling (TS) and safety sensing and signaling (SS), along with more fully understanding their respective systemic responses and influences.

Cytokines are a class of small molecule proteins or polypeptides that are stimulated, synthesized, or secreted by cells. The cytokine network is a large family (more being discovered every year) of peptides, roughly 100-250 amino acids in sequences folded into complex forms. Via apocrine, paracrine, and endocrine mechanics, they coordinate systemic responses to the environment. This results in not only mitochondria, cell, tissue, and organ responses, but emotional, behavioral, social, and mental responses as well.

Cytokines transmit information between cells. Cytokines have immune regulation and effector functions, including the regulation of innate immunity, adaptive immunity, surveillance immunity, cell growth, cell differentiation, and repair of damaged tissues. Cytokines play an important role in the immune system by regulating the intensity and duration of immune responses.

Many cytokines promote or restrict each other, forming an extremely complex regulatory network that has various properties, such as pleiotropic and overlapping agonist and antagonist activities. Elevation of proinflammatory and catabolic cytokines leads to the occurrence of disease, disability, pain, and suffering, and extreme elevation leads to cytokine storms and cytokine storm syndrome, resulting in shock, multiple tissue and organ damage, functional failure, and death.

Complex and dynamic interactions between cytokines and their receptors mediate immunity (defense), inflammation (destruction), and metabolism (fuel). Cytokines are the final effector proteins involved in auto-immuno-inflammo-metabolic diseases. Associated cytokine mediated phenotypic changes can even change the immunosuppressive tumor microenvironment, leading to tumor initiation, progression, and metastasis. With the improvement of biopharmaceutical technology, there is a vast growth space for targeted cytokine drugs in the future, which is a promising prospect for the treatment of tumors, auto-immuno-inflammo-metabolic diseases, as well as other diseases, and associated disability, pain, and suffering.

There are resources available to understand the more specific and nuanced functions of each cytokine.[94] However, as cytokines tend to have relative specificity to threat versus safety, for the purposes of explanation over precision, the cytokines will be lumped into two large categories—*Threat Cytokines* (TC) and *Safety Cytokines* (SC). [95]

Threat Cytokines (Examples: Tumor Necrosis Factor alpha (TNFa), Interleukins: IL1b, IL2, IL6, IL8, IL17, IL23)

The cytokine network is a major intercellular signaler of threat and an excellent network to follow to understand TS and the associated phenotypic and physiologic changes associated with threat.

- TC help coordinate a systemic effort to thwart threats and survive attacks.[96]
- TC regulate phenotypic expression in mitochondria, cells, tissues, organs, and networks (immunologic, hematologic, endocrinologic, neurologic, autonomic, fascio-musculo-skeletal, cardiovascular, respiratory, gastrointestinal, genitourinary, adipose, fibrous, osseous....) of the body towards a coordinated defensive strategy.
- TC have most notably been associated with the activation of inflammatory pathways and the immune response. However, the TCs have many other functions that help signal a systemic response to threat.[97]

- TC influence the methylation of DNA, modifications of histones, and adjustments to microRNA to change genomic expression—threat-associated epigenetic modifications.[98]
- TC signal transcription factors to induce transcription, phenotypic transitions, and phenotypes of threat.[99]
- TC loosen intercellular tight junctions to help the system battle invasive organisms within the interstitial spaces.[100]
- TC increase coagulation factors and platelet functions to stop bleeding and to trap invasive organisms when under threat.[101]
- TC switch both mitochondrial and cell phenotypes to more defensive, inflammatory, and aggressive forms.[102]
- TC can signal cells to deconstruct, recycle, and repurpose (autophagy).[103]
- TC can signal cells to engage in different programs for cell death—apoptosis, necroptosis, and pyroptosis.[104,105]
- TC induce transaminases, enzymes for glutamate production, and amplify glutamate production and signaling at glutamate receptors (glutamate, an ancient monoamine, is another major intercellular signaler involved in a threat response).[106,107,108,109]
- TC block the production of other monoamines, acetylcholine, and peptides used in safety-signaling pathways.[110,111]
- TC regulate cell receptors and neuronal synapses to support a TR.
- TC alter the autonomic nervous network to support fight and flight, as well as to induce falter and faint physiologies. Within this context, they are also responsible for shock syndromes.[112]

- TC alter cortical and subcortical brain functions to support a TR.[113]
- TC alter the expression of the hypothalamic-pituitary-adrenal network to support inflammation, blood volume, fuel production, and an overall TR. [114,115]
- TC alter emotional feelings, instinctive behaviors, thought bias, social interactions, and mental processes, to support a TR. [116,117,118,119,120,121,122,123, 124,125]
- TC also dictate fuel selection and metabolism through complex pathways.[126,127]
- TC activate threat-related metabolic states and fuel production (glucose) by inducing a strategic and organized breakdown of tissues for the fight, flight, and fever states, as well as the fever state of an infection.
- TC induce insulin resistance in nonessential cells, tissues, and organs for fight, flight, and fever to preserve glucose for their essential counterparts.[128]
- TC induce glucagon production, glycolysis, glycogenolysis, and gluconeogenesis.
- TCs induce proteolysis and lipolysis.
- TC induce inflammatory low-density lipoprotein-cholesterol formation and reduce anti-inflammatory, high-density lipoprotein-cholesterol formation and conversion.[129]
- TC reduce beta oxidation of fats.[130]
- TC induce toxic reactive oxygen species production. [131]
- TC shift mitochondrial and cellular functions away from the high-energy production, anti-inflammatory, and anabolic metabolic state of oxidative phosphorylation to the faster, yet less efficient, defensive, inflammatory, oxidizing, and catabolic metabolic state of glycolysis.[132]

It is important to consider that the shift in metabolism to glycolysis produces energy at a faster rate than oxidative phosphorylation which is needed for fight, flight, and fever. Yet, it results in substantially less energy per molecule of substrate to run the functions of the cell while mounting a defense. In a TR, there is a dissolution to this more primitive metabolic strategy, perhaps specifically for rapid production of energy through glycolysis. (Glycolysis predates the evolution of oxidative phosphorylation by billions of years.)

To sustain this glycolytic physiology, there is a demand for much more substrate. Yet the organism, when in a defensive state, is not able to acquire or accumulate substrates. There is no time for lunch when you are about to *be* lunch. Therefore, the system is relegated to using internal resources for substrate—catabolism. This high-energy demand, low substrate available, and low-energy efficiency state is not sustainable for very long.

TC are catabolic—the classic threat cytokine tumor necrosis factor alpha (TNFa) was first known as a hormone, cachectin, because it was noted to cause cachexia (wasting).[133] At a micro level TNFa signals mitochondrial and cellular changes to induce glycolysis, insulin resistance, inflammation, oxidation, and coagulation. TNFa is also a signaler of emotional, behavioral, social, and mental changes and is associated with anxiety, depression, and higher-level cognitive impairment.

When in threat, the oxygen spared from mitochondrial oxidative phosphorylation formation is diverted to the mitochondrial and cellular production of toxic and inflammatory reactive oxygen species and oxidative enzymes to aid in a defense from a pathogen or predator. It is notable that mitochondria have multiple phenotypes, as do cells… and humans. Mitochondria under the influence of TS and TC undergo fission to divide and scatter away from the cell's organelles to the periphery of the cell, assuming a defensive posture. Too often, mitochondrial "dysfunction" or "exhaustion" is used to describe mitochondria as defective when the mitochondria are doing exactly what they are supposed to do given the environment of threat present. Similarly, humans are described as defective or pathological when they are doing exactly what they evolved to do considering their own environment of threat.

When in threat, fats do not easily undergo beta oxidation for energy production. They can be diverted or converted from anabolic and sex hormone production, oxidative fuel production, and cell structure construction to distress hormone and inflammatory molecule production.

Interestingly, lipids within the mitochondria and cell membranes change structure to become more rigid and less penetrable within threat phenotypes.

When threatened, the TCs alter the autonomic nervous system for various defense strategies. The anti-inflammatory, anabolic, autonomic parasympathetic network for bond, breed, feed, digest, nest, and rest is offline in threat states.

Within the above bullets, one can see how the catabolic and inflammatory physiology of a TR left unchecked contains all the underpinnings for the formation and sustenance of chronic diseases and disability, along with pain and suffering.

Safety Cytokines (Transforming Growth Factor beta (TGFb), Brain Derived Neurotrophic Factor (BDNF), Leukemia Inhibitory Factor (LIF), Granulocyte Colony-Stimulating Factor (G-CSF), Irisin, Interleukins: IL-4, IL-10, IL-11, IL-13, IL-19, IL-35, IL37)

The cytokine network is a major intercellular signaler of safety. It is an excellent network to follow to understand SS and the phenotypic and physiologic changes associated with safety, in contrast to the changes associated with threat.

- SC help coordinate a systemic effort that allow humans to thrive.[134]

- SC regulate phenotypic expression in mitochondria, cells, tissues, organs, and networks (immunologic, hematologic, endocrinologic, neurologic, autonomic, fascio-musculo-skeletal, cardiovascular, respiratory, gastrointestinal, genitourinary, adipose, fibrous, osseous....) of the body towards a coordinated construction and reproduction strategy.
- SC have most notably been associated with deactivation of inflammatory pathways and innate immunity, along with increased surveillance immunity and healing. However, the SCs have many other functions that help signal a systemic response when the world is perceived as safe.[135]
- SC influence the methylation of DNA, modifications of histones, and adjustments to microRNA to change genomic expression—safety-associated epigenetic modifications.[136]
- SC signal the transcription, phenotypic transitions, and phenotypes of safety.[137]
- SC signal cell growth, regeneration, reproduction, recovery, and healing.[138]
- SC influence programmed cell deconstruction and recycling (autophagy).
- SC influence programmed cell death—apoptosis.[139]
- SC induce the production of feel-good monoamines such as dopamine, serotonin, and melatonin.
- SC influence the production of the neurotransmitter acetylcholine
- SC indirectly increase the peptide hormone oxytocin, as well as steroidal sex hormone production.[140,141,142]
- SC regulate cell and neuronal synapse receptors to support a SR.
- SC alter the autonomic nervous network to support

the anti-inflammatory and anabolic parasympathetic functions of bond, breed, feed, digest, nest, and rest.

- SC alter cortical and subcortical brain functions to support healing and vitality.[143]
- SC alter the expression of the hypothalamic-pituitary-adrenal network to support appetite, growth, repair, regeneration, and reproduction.[144]
- SC alter emotions, social interactions, mental processes, and behaviors to support a SR.[145]
- SC also dictate fuel choices and metabolism through complex pathways.[146,147]
- SC change metabolic states and fuel production to become more anabolic and efficient.[148,149]
- SC induce insulin sensitivity.[150]
- SC induce anti-inflammatory, high-density lipoprotein-cholesterol formation and reduce inflammatory, low-density lipoprotein-cholesterol formation and conversion.[151]
- SC facilitate beta oxidation of fat.[152]
- SC shift mitochondrial and cellular functions to the high-efficiency, high-energy, anti-inflammatory, antioxidant, anabolic, and healing metabolic state of oxidative phosphorylation. They are shifted away from the less efficient, defensive, inflammatory, and catabolic metabolic state of glycolysis.[153]

When in safety, through mitochondrial oxidative phosphorylation, cells are capable of up to eighteen-fold more energy production per substrate than when in threat through glycolysis. SS results in mitochondrial transformation and fusion to a phenotype with increased mitochondrial oxidative phosphorylation capacity.[154] Mitochondria undergo fusion when safe. They move away from their defensive posts at the periphery of the cell back to the organelles, where they produce and supply the energy needed for the anabolic and reproductive functions of the cell.

When safe, oxygen utilization for toxic enzymes and inflammatory reactive oxygen species is markedly reduced. Oxygen is used for high energy production through oxidative phosphorylation.

When safe, fats are used for anabolic and sex hormone production, oxidative fuel production, and cell structure construction. Meanwhile, the production of distress hormones and inflammatory molecules is reduced.

When safe, the anti-inflammatory, antioxidant, anabolic, autonomic parasympathetic network for bond, breed, feed, digest, nest, and rest is online.[155]

Humans must get to this state of safety to fully heal and be resilient. This end of the threat-safety spectrum is where healing, health, wellness, and thriving lie.

"The more we discover that we humans
are fantastically complicated creatures.
Yet, on the other hand, human life is full of moments
when we know that things are also very simple."

Tom Wright

It is probable that the pleiotropic—and sometimes perceived contradictory or paradoxical—nature of the cytokine signaling system is secondary to the folded configurations of these peptides, ligand concentrations, receptor types, receptor affinities, receptor densities, receptor solubilities, cofactors, and signaling adaptors of the different cells, tissues, and organs of the body.

Simple examples of pleiotropic mechanisms exist within monoamine and steroid hormone communication networks. The monoamine network is ancient and present in bacteria.

This network has been conserved and advanced over time. The monoamines are simpler in structure than the cytokines. Phylogenetically, the monoamines precede the cytokines, likely, by billions of years. The steroid hormone network is phylogenetically newer than the cytokine network. Evolution of the cytokine network began approximately 2 billion years ago. The steroid hormone network started evolving along with multicellular, differentiated-cellular, and large life approximately 500 million years ago. As vascular networks assembled to communicate and coordinate the cells across these larger systems, hormones evolved within differentiated glandular cells, tissues, and organs. These ligands were released into the blood for transport across the system in search of their receptors. Although, there are differences in the age and structure of these networks, when in search for their receptors, the monoamines and steroid hormones use similar mechanisms to account for their pleiotropic effects. The cytokines very likely use similar and perhaps more complex mechanisms to account for their pleiotropy—many of these mechanisms are yet to be discovered.

Noradrenaline, a monoamine, in low concentrations sticks to high affinity receptors that lower heart rate and blood pressure and provide a state of attention and calmness. Medium concentrations of noradrenaline stick to medium and high-affinity receptors to elevate heart rate and blood pressure, and to give a state of arousal, cognitive acuity, and high physical, mental and sexual performance. High concentrations of noradrenaline saturate high, medium, and low-affinity receptors. These low-affinity receptors further increase heart rate and blood pressure and provide a state consistent with fight and flight (T1) physiology. This spectrum includes anger to fear, as well as impaired neocortical, gastrointestinal, and reproductive functions. In this state, frequent complaints include inability to think, inability to speak, and being "out of one's mind." This state can also be accompanied by asocial, if not antisocial, behavior.

Cortisol, an adrenal cortex steroid hormone, in low concentrations sticks to high-affinity mineralocorticoid receptors and increases inflammation and acute immunity physiology. Cortisol in high concentrations saturates both the high-affinity mineralocorticoid receptors, and now the low-affinity glucocorticoid receptors, to reduce inflammation and acute immunity physiology. In addition, threat-induced hypomethylation (*a decrease in the epigenetic methylation of DNA)* of cortisol-related genes increases the sensitivity of cortisol receptors—an epigenetic change. Acutely, supraphysiologic boluses of cortisol can induce hyperarousal, and even mania. By contrast, chronically, supraphysiologic cortisol will induce depressed immunity, mood, and cognition, and eventually, induce neocortical atrophy and degeneration.

Single ligands with multiple effects are based on their configurations, concentrations, receptor affinities and modifications, and cell, tissue, and organ specificity. This is how biphasic, perhaps better stated as polyphasic, but not contradictory, phenotypes and physiologies can be created by a single ligand.

The cytokines likely have similar mechanisms at play to account for their pleiotropism. The TC thus have the ability based on concentration to take human physiology from fortress to fight to flight to fever to freeze to falter to faint. In an acute infection, such as Covid-19, this escalating process can play out over days, culminating in a "cytokine storm," with shock, extensive cell, tissue, and organ damage, and even death. In a chronic state, this process can play out slowly, yet be unrelenting, culminating in chronic disease, disability, pain, suffering, and premature death. Mental illness, cardiovascular disease, fascio-musculo-skeletal degeneration, neurologic degeneration, cancer, premature aging, and again, even early death can ensue from chronic threat.

This simple ligand, along with receptor, cell, tissue, and organ model can explain away a lot of confusing complexity.

Another contributor to the complexity of the system is context—mostly whether an organism is in threat or safety within its environment. It is important to know the context to understand intention in the phenotypic and physiologic changes and presentations of an organism. Without this fundamental understanding, data can become confusing, if not nonsensical. Although genomes are relatively simple in structure, they are somewhat plastic, adding even more complexity. But more importantly the exposome is infinitely complex. It is this interface of the genome and exposome that gives life its context and ultimate complexity. Context is missing from most research protocols, medical models, and treatment algorithms.

It will be important to build models of disease, disability, pain, and suffering that can explain how a single signaler can induce multiple phenotypic changes within an organism along a spectrum from threat to safety. This may seem complicated. However, through a lens of configurations and concentrations, receptor types, affinities, and modifications, cofactors adaptors, and known context, this apparent complexity may become transparent, purposeful, and explainable. Ultimately, any current confusion around cytokine pleiotropic effects may prove to be neither contradictory nor paradoxical at all.[156]

Clinical Considerations and Implications of TVST in Disease, Disability, Pain, and Suffering

> *"Without health life is not life; it is only a state of languor and suffering – an image of death."*
>
> *Buddha*

Why are human illness, disease, disability, pain, and suffering on the rise, while human brain volume, intellect, and lifespans are falling within the United States despite advances in medical knowledge and technologies?[157] The answer to this lies in TVST—individual, societal, and global TLs are climbing. With increased threat, humans are not only more inflamed, oxidative, catabolic, and degenerative, but immune functions are altered to be either hyperactive or hypoactive. The variability of the threat-self depends on the expressed T1 versus T2 phenotypes. Additionally, TS induces the expression of genetic changes and oncogenomic clusters, resulting in cancer cell phenotypes.[158] These dynamics underpin chronic disease, disability, pain, and suffering.

Of note, these dynamics are a coordinated, systemic response to threat, of which inflammation is just a part.[159] Threat-induced catabolism, along with emotional, behavioral, social, and mental changes, may be the more important components of chronic disease, disability, pain, and suffering that are infrequently incorporated into models of pathology, or perhaps models of phenology, and care.

The Latin terms for describing inflammation include *rubor* (redness), *calor* (heat), *dolor* (pain), *tumor* (growth/swelling), and *fluor* (secretion). These terms date back to at least the 14th century, along with the term *humor* (fluid). Humor refers to the ancient Greeks' idea of four main fluids that were felt to form and flow within human bodies—*blood, phlegm, yellow bile, and black bile*. It was felt these fluids determined emotions, behaviors, and temperament. This concept is inaccurate in detail, but somewhat correct in concept. It is the interstitial fluid that baths the cells that determines so much, including inflammation of the cells and tissues, but also more broadly the metabolism, connections, emotions, behaviors, thoughts, and relationships of the organism. Good *humor* comes from the bottom up—the interstitial fluid meeting the cells while in safety.

"The body of man has in itself blood, phlegm, yellow bile and black bile; these make up the nature of this body, and through these he feels pain or enjoys health. Now he enjoys the most perfect health when these elements are duly proportioned to one another in respect of compounding, power and bulk, and when they are perfectly mingled."

Hippocrates

It is important to have the perspective in chronic disease, disability, pain, and suffering that the system is doing exactly what it is supposed to do given the TL. It isn't broken, just different. The outcome is unwanted, and therefore, maladaptive. The problem is the cumulative, compounded, chronic, and unrelenting TL that the system is responding to—4.2 billion years of genetic evolution and programming, interfaced with an environment of threat playing the genetic code exactly how it was designed to be played.

When the system is activated in a TR, the expression of the genome, nucleosome, transcriptome, proteome, metabolome, inflammasome, connectome, biome...the phenome, all change—emotional, behavioral, cognitive, sociologic, psychologic, metabolic, endocrinologic, immunologic, neurologic, cardiovascular, respiratory, gastrointestinal, genitourinary, adipose, muscle, fibrous, osseous, mitochondrial, oxidative, inflammatory, and cellular statuses change. Physiology changes. The system's status changes.

The attribution of disease to inflammation alone is inappropriate. Inflammation only represents a fraction of the physiology of threat. Inflammation in a systemic TR correlates with chronic disease but is not the sole cause of chronic disease. Threat is the source and cause, and inflammation is part of the TR.

Additionally, assessing physical threats to understand this phenomenon is inadequate. Analyzing generational, historical, ecological, emotional, behavioral, social, mental, cultural, societal, institutional, technological, financial, and spiritual threats is required to completely evaluate the TL of a human being. Elevated threat, TS, and TC are associated with all chronic emotional, behavioral, social, mental, and spiritual diseases, disability, pain, and suffering, including addictions. [160,161,162]

It needs to be reinforced that both the interosome and exterosome can contain threat and initiate TS. This, in turn, activates a mitochondrial, cellular, tissue, organ, emotional, behavioral, social, mental, and spiritual response—a systemic TR. It is vital to recognize that the nonphysical threat-related illness and disease burden usually does not have its origin in a toxin, pollutant, mutagen, allergen, antigen, pathogen, or physical injury. Nor does this illness and disease burden come from genetic, brain, or character deficiencies or defects. This burden comes directly from the TL of an individual and/or society, and reciprocally, the lack of safety in the world.

Threat causes the greater system to prioritize, and resource allocate. Meanwhile, the mitochondria, cells, tissues, and organs specialize in function, specifically for the defensive tasks at hand. Every mitochondrion, cell, tissue, and organ in the body changes its expression and function under threat. Some increase activity, some decrease activity, and some even sacrifice themselves for the greater good.

The endocrinologic, immunologic, neurologic, and autonomic networks all change their status in threat. [163,164,165,166] The functions of the respiratory, cardiovascular, and muscular networks are prioritized and up-regulated in early threat. The functions of the reproductive, genitourinary, gastrointestinal, fibrous, and osseous networks are down-regulated in threat. [167,168,169,170,171]

For example, the pancreas, an endocrinologic gastro-intestinal organ, (perhaps neurologic and immunologic organ, too[172]) serves as a relatively simple organ to illustrate how things change under threat. Look deeper than the organ, tissue, cells, and mitochondria to the level of the genome and the transcription and translation of the genetic code to fully appreciate the mechanisms of a TR.

TS induces certain threat-associated transcription factors and inhibitors within the various cells of the pancreas. These decrease expression of insulin-producing beta cells and increase the expression of glucagon-producing alpha cells. This results in the decreased production of insulin and the increased production of glucagon, with the goal to mobilize, and not store, glucose—the needed fuel for the fight, flight, and fever. TS turn down the pancreas' production and secretion of digestive enzymes. This allows resources and energy to be utilized elsewhere for the fight, flight, or fever.

The pancreas is a colony of many different, differentiated, and specialized cells within the same organ. These cells respond differently to the many different environmental stimuli within a well-coordinated systemic response—including threat.

Another, but more complex example is the brain, a neurologic, endocrinologic, and immunologic organ. The brain can be considered as a colony of different, differentiated, and specialized cells within the same organ that change expression within a defense response. Parts of the newer human brain, the neocortex—responsible for social connection, created ideations, cognitive constructs, symbolic narratives, declarative memory, and a significant amount of energy consumption—are not very useful in a pathogenic or predatory fight, flight, or fever response. Bonding, creativity, calculus, poetry, and playing Jeopardy just don't help much in a tiger fight. Therefore, these parts of the brain get turned down. Meanwhile, the more primitive, faster reacting brain structures, such as the sensory, striatal, and subcortical limbic circuits are prioritized and enhanced. In addition, the parasympathetic autonomic functions are turned down when in threat. The sympathetic autonomic functions are turned up in early threat.

The immunologic, neuromodulating, and connective tissue cells—glial cells—within the brain change expression in threat. Neuronal functions, including transmission, are regulated to provide optimal signaling to survive a threat. The hypothalamic-pituitary-adrenal network changes expression in threat as well.

All these changes can be automatically signaled within a TR. The initiation of this signaling can come from the first cell to sense threat. Whether that cell is in a toe or in the nose, the system will be placed on alert and adjusted in a threat response. A threat response can occur prior to, and even in the absence of, cortical awareness. Humans have some cortical cognitive capabilities to modify this response, but much less than most might imagine. Restoration of safety is the key to controlling the TR and recovery.

What do humans look like emotionally, behaviorally, socially, mentally, and spiritually when these threat-induced changes happen?

Threat, TS, TC within a TR notably impair contemplation, rationalization, organization, creation, connection, and socialization. In threat, humans turn up the older and more primitive parts of the brain that facilitate competition, aggression, reaction, mobilization, and protection. Threat notably can induce anger, fear, sadness, despair, impulsivity, vigilance, paranoia, prejudice, disorganization, disconnection, and asocial, if not antisocial, behaviors. If this pattern persists, what once was acutely adaptive, in chronicity, may become maladaptive. Adverse outcomes precipitated from threat behaviors may additionally increase the TL. The result is a downward spiral of dysfunction, disease, disability, along with pain and suffering.

> *"Mental health problems don't define who you are.*
> *They are something you experience.*
> *You walk in the rain and you feel the rain,*
> *but you are not the rain."*
>
> *Matt Haig*

Historically, mental health care has been focused on the description and classification of symptoms, rather than on underlying causes. The Diagnostic and Statistical Manual of Mental Disorders contains descriptions, symptoms, and diagnostic criteria. The DSM has grouped patients into cohorts without any sense of the underlying mechanisms behind their conditions, or for that matter simply without any sense. Attempts to find causal mechanisms rather than descriptive correlates for mental illness conditions have been unsuccessful. The National Institute of Mental Health has recently provided funding for basic research on disease processes of the brain. Over $20 billion of funding failed to make a dent in mental illness care. Biomarkers and genes were identified in mental illness conditions but were of little clinical significance. Genes alone are clearly not the answer to the mental illness pandemic.

The United Kingdom Biobank recently published data revealing that people with depressive episodes had significantly higher levels of inflammatory proteins, such as TC, in the blood. Inflammation, immune functions, and autoantibodies are of high interest in mental illness research today. The newest rage. Work on patients with immune-driven psychosis suggests that a range of strategies including removing antibodies, immunotherapy drugs, or steroids can be effective treatments. As well, metabolic disturbances have also been identified in mental illness conditions including psychosis, bipolar disorder, major depressive disorder, and eating disorders. Interestingly, a ketogenic diet, in which carbohydrate intake is limited, may be of benefit as ketone bodies suppress transcription factors and the production of TC.

Furthermore, many mental health experts and treatment guidelines operate on the assumption that individuals have a single principal mental health diagnosis, yet medicine is becoming more aware that serious mental illness, including the likes of bipolar disorder or schizophrenia spectrum disorders, is associated with an increased risk for many comorbid physical illnesses. A recent study demonstrated this increased risk to be twofold. Physical comorbidities, which included cardiovascular, endocrine, neurological, gastrointestinal, musculoskeletal, and infectious disorders, are prevalent at similar rates in both bipolar and schizophrenia spectrum disorders. Additionally, mental illness is associated with reduced life expectancy.

Experts have stated that additional chronic illnesses—whether physical or psychiatric—may underlie these associations. These experts also speculate that physical comorbidities within mental illness could stem from side effects of psychotropic medications, which are known to cause cardiometabolic changes, including weight gain, or lifestyle factors and non-modifiable risk factors could also contribute to physical comorbidities. The experts also suggest that a greater understanding of the epidemiological manifestations of comorbidities in mental illness is "imperative." Yet, treatment of physical and mental illnesses remains siloed.

TVST differs in perspective and suggests that there should be little angst and confusion around mental illness and its causes and correlates—that this is all very understandable and mental illness, and the associated comorbidities, all arises from the same soup from the same root—chronic threat. What can be agreed upon is that the high prevalence of physical comorbidities demand urgent attention to the integration of care models that address both physical and mental health outcomes in people with severe mental illness. TVST proposes this should be done with a focus on reducing chronic threats.

In addition, what is missing here is the realization that genes, immune responses, metabolic responses, and inflammation are not the causes of mental illness...just the correlates. All these things are major components of a TR. Chronic threat is the cause of mental illness. It seems more than reasonable to stop wasting money on the manipulating correlates and to start treating the cause. Increasing "awareness of mental health" and hanging a diagnosis around someone's neck does little to actually treat a chronic threat condition. A new paradigm and campaign is needed to treat at the root—chronic threat.

Psychology is physiology—there is no mind-body duality. Psychological pathology is programmed chronic, cumulative, compounded, and yet maladaptive threat physiology.[173] It all comes from the root and soup of threat. This physiologic soup can account for the escalation in emotional, behavioral, social, mental, spiritual, and cultural disease, disability, pain suffering, and despair seen in the world today.[174] Primary psychological and cognitive disorders arise directly from chronically sustained states of threat, TS, TC, and the threat-associated changes in the physiology and the connectome of the brain.

The connectome references the variable connection patterns of the brain—the sum of the different connectotypes of the brain—which reflect the exposome-genome interface, and subsequent phenotype and physiology. The brain has hubs and networks, and although everything is connected through the physiologic soup, not all brain neuroconnections are connected all the time. At different times, different parts of the brain are downregulated, disconnected, or dissociated, while other parts are upregulated, connected, and aware. To be otherwise would lack specificity for needs, consume excessive energy, and result in chaos.

Adaptive physiology, emotions, and behaviors for a defense from acute threats that are sustained chronically when suffering from unrelenting threat become maladaptive within the chronic connectotype of threat physiology. Anxiety and depression are not monoamine deficiencies, nor broken brain pathological states, but maladaptive physiological states of chronic threat from deficient and broken lives.

Downregulation, disconnection, and dissociation can advance to destruction and degeneration of brain structures. Yet, the system itself is doing exactly what it is supposed to when in threat. The system is neither broken nor pathological. It is the chronic toxicity of the exposome interfacing with the genome that produces these phenotypes of threat. Chronic threat and the chronicity of this toxicity is the source of dysfunction.

Within current disease models, "symptoms" are targeted to try to cure the illness. Although in their chronicity, the phenotypes may be maladaptive, targeting these "symptoms" depletes defenses, leaving organisms more vulnerable to threats. This is a bad strategy, a bad paradigm for healing, health, and wellness. In getting to the root of the symptoms, the total TL must be evaluated and treated. Treating the symptoms does not cure the disease.

Within chronic threat, the T1b phenotype exhibits chronic irritability, anger, opposition, rage, explosivity, and impulsivity disorders. Whereas the T1c phenotype exhibits chronic fear, anxiety, hypervigilance, and paranoid disorders. The T2b phenotype exhibits chronic sadness, depression, hopelessness, and avoidance disorders. And the T2c phenotype exhibits chronic despair, numbing, helplessness, and severe dissociative disorders.[175,176,177, 178,179] Homicide and suicide follow these threat phenotypes.

Attention deficit disorder, hyperactivity disorder, bipolar affective disorder, post-traumatic stress disorder, autism spectrum disorder, conversion disorder, and even schizophrenia fall into this spectrum of primary psychological, cognitive, and behavioral disorders. All of these disorders are associated with elevated threat, TS and TC, and threat-associated changes within the connections and functions of the brain, plus the potential of premature neurodegeneration, as well as many other threat-related comorbidities.[180,181,182,183,184, 185,186,187,188] Attention deficit disorders, autism, and psychoses all share changes in the functions of the insular cortex (housing interoception, sense of self, and beliefs), the ventral striatum (responsible for emotion, motion, and motivation), and the salience network (which orients attention primarily to external stimuli). Autism is associated with three times the rate of Parkinson's disease—same soup.

Autism notably is a phenotypic change in brain functions associated with hypersensitivities; emotional, behavioral, social, and mental disturbances; altered cognitive acumen; depleted language production; and atypical memory consolidation and retrieval processes that look suspicious for threat bias and threat physiology.

Attention deficit hyperactivity disorder is associated with premature cognitive decline, Lewy Body formation, and dementia later in life. The same threat root soup simmering over a lifetime.

Conversion disorders are severe threat and trauma induced phenotypic and physiologic changes associated with things like numbness, blindness, paralysis, and fits without any associated pathology.

Neurosis, obsessive compulsive disorder, bipolar affective disorder, and hypochondriasis have been associated with an increase in all-cause mortality. Neuropsychiatric disorders in general share a substantial and largely overlapping imprint of poor body health consistent with a chronic TR.[189]

Humans need to be safe, seen, and secure to feel and heal.

"The boy's life starts to represent the whole nature of humanity – we all have this self-imposed deaf, dumb, and blindness..."

Pete Townsend discusses Tommy

Both ecological and psychological toxins and pollutants place humans at risk for the same set of neuropsychological and physical diseases. Air pollution and abuse stimulate very similar physiologic response. Lipopolysaccharide-stimulated (PAMP-stimulated) immune-inflammatory markers, i.e., LPS-stimulated TS and TC are upregulated and elevated in proportion to childhood trauma severity. These TC levels are more affected by experiences, such as childhood trauma, than lifestyle or behavioral choices.[190] Adverse childhood events (ACEs) induce coding changes and a hyperactive TR that lead to increased physical and mental illness across a lifetime—if not decoded and recoded.

Primary psychological diagnoses are associated with elevated distress related inflammatory, oxidative, and metabolic biomarkers—CRP, TC, glutamate, glycolysis, N-glycosylation, LDLs...and mitochondrial dysfunction.[191] All psychological disorders are associated with "sapiocortical dysfunction" and threat states induce more mitochondrial DNA insertions into the genomes of the cells of the "sapiocortext" than into any other cells. Psychological disorders are the cascade of threat physiology extending from the mitochondria to DNA/RNA, glutamate, monoamine, peptide, and hormone signaling, to blood flow and neuroelectrical currents, to the microbiome, all designed in some way to help us survive the threat. The TR is not the problem the TL and its severity and chronicity are the problem.

"I know you are busy pulling the wool over my eyes but you may want to address the inconsistencies in your persona first."

Et Imperatrix Noctem

Secondary psychological diagnoses include personality, distraction, compulsion, and addiction disorders.

TVST defines personality disorders as the *chronic maladaptive use of illusions of control, constraint, repression, and suppression strategies.* False-self, ego, protection offer an illusion of control. Yet, this can create internal conflict by constraining, repressing, and suppressing emotions, behaviors, and thoughts. Constraining, repressing, and suppressing can be helpful to negotiate acute threats in the short run. However, these same mechanisms can become maladaptive over the long run and create more dysfunction, conflict, and threat. This, in turn, leads to more disease, disability, pain and suffering.[192] Indeed, inhibition—constraint, repression, suppression—correlate with elevated TC and threat associated changes in brain physiology and the connectome that are consistent with other patterns of threat.[193,194]

Distractions and compulsions prevent dealing with the physiology bubbling up from below and are forms of repression and suppression mechanics that can perpetuate or increase internal conflict. Intoxicants numb both physical and emotional pain temporarily, but lead to increased threat physiology chronically, a formula for addiction. Just as substance use is an analgesic for emotional, social, mental, and spiritual pain and suffering, so too are illusions of control analgesics for emotional, social, mental, and spiritual pain and suffering. Long-term use of both can be maladaptive and result in more pain and suffering, as well as illness and disease.

Overall, these maladaptive strategies, used in excess, will

cause chronic internal and external conflict, increasing TS, and thus preventing the finding of safety in the world.

Psychological therapy needs to keep these principles in mind when creating a logical, linear, efficient, and progressive treatment algorithm. The reliance on the transformation of somatoemotional awareness into a cognitive construct—Name it to Tame it—prior to free emotional expression—Feel it to Heal it—and safe social connection may be a backwards and less effective approach. It would be prudent to first provide for the freedom of emotional expression in a safe and secure setting prior to loading the chronically threatened with cognitive constructs. Talk therapy may not be the place to start mental health treatments. Patients may not be physiologically ready for such therapy. As cognitive loading increases TS, TC, and the TL, cognitive work misapplied or mistimed may make symptoms worse. This can be counterproductive to healing, health, and wellness, particularly in those who already have extremely high TLs.

Threat biases away from higher level cognition and the present moment. In threat, humans are reactive and not contemplative while below cortical awareness the system is trying to use past codes to predict the future. This is human physiology—this is biology. To try and force the construct of what it looks like to be safe—cognitive and present—onto someone in threat—reactive and predictive—is not a cure and only creates an illusion of improvement. This can increase internal conflict and activate the TR further.

Biophysiologically, health seems more predicated on the safe expression of emotions, behaviors, and thoughts—the natural flow of energy—than on cortical or cognitive awareness of the problems. Social contracts, created ideations, cognitive constructs, symbolic narratives, and rigid beliefs can be barriers to this flow. Some of these higher human cortical functions may be adaptive in the short run. However, in the long run, they can become maladaptive to human survival and the propagation of the genetic code.

To have awareness and to understand the relationship of physiologies and drives to emotions, feelings, behaviors, and thoughts has some validity to prevent or deconstruct maladaptive strategies and habits. However, language-based ideation, cognition, and narration do not seem to be primary in the 4-billion-year-old processes of healing physiology.

"[Threat] is the wind that blows out
the lamp of the mind"

to paraphrase Robert Green Ingerson.

The phenotypic changes and dissociations of threat reduce both emotional and intellectual quotients. Threat load is not just physiologically altering the modern human brain, but physically and structurally shrinking it.[195] Safety restores.

Children are particularly suffering from both mental health disorders and learning difficulties. They can additionally be burdened by the weight of these many diagnoses. More than 1 in 10 individuals between ages five and twenty-four live with at least one learning and/or mental health diagnoses (and the incidence of these diagnoses doubles between the ages of five to twenty-four years old) that all come from the same root—threat.[196]

Finland is a model for emotional, intellectual, and educational reform. By reducing threat and increasing safety across their schools, society, and culture, their children are outperforming the world—most notably children in the U.S. The Finns' literacy rate is 99.0%, primary school enrollment is 99.7%, and secondary enrollment in academic and trade schools is 66.2%. 93.0% of Finns graduate from high school, 17.5 percentage points higher than the United States. Yet, Finland spends about 30% less money per student than the United States.

The Finnish educational system has shunned standardized testing, given teachers creative freedom, focused on cooperative, not competitive learning, and prioritized the basics—the foundations of learning. In addition, they start formal schooling at a later age and begin school days later to allow for restorative sleep. The academic atmosphere is more relaxed and conducive to collaboration and creativity. The Finns have little homework, and time for art and play is preserved. Finally, the Finnish educational system has many more funded options for vocational training, beyond traditional college education. Labor and the trades are valued.

The Finns have applied the principles of TVST in education and schools to great success.[197] The Finns have focused on growing better brains and better human beings within a culture of cooperation and safety, rather than improving SAT scores within a culture of competition and threat.

The United States needs an algorithm of wellness and health like the Finnish educational system. The U.S. needs to focus first on the nurturing and healing processes—cooperation, collaboration, adaptation, and expression, and then the cognitive processes—awareness, attention, reformation, and consolidation. Additionally, the U.S. needs to back away from the toxic pushing of individual potential through competition with the rest of the species and embrace an algorithm that potentiates the species as a whole from which the potential of all its individuals naturally arises.

Similarly, ecological, physical, emotional, social, cultural, and financial safety should precede cognitive processes and mental therapy as much as possible. Proper prioritization and sequencing of both educational and mental health algorithms—from the code to the cell to the subcortex to the cortex—should result in faster and better outcomes in emotional, behavioral, social, mental, intellectual, and spiritual health care.

"You are not seeing reality. You are seeing what you are believing. Stop believing and you will see more reality."

Shree Swami Premodaya

To reiterate, there is no mind-body separation. Those with mental health issues also have more physical, behavioral, social, cognitive, spiritual, and learning issues. They die at a younger age, primarily from physical health issues, such as strokes and heart attacks. It is all the same soup from the same root—threat.[198]

Cerebrovascular, neurodegenerative, movement related, psychiatric, developmental, learning, and substance abuse disorders, as well as brain tumors and other brain-related diseases all have commonalities. Except for diseases arising from specific single-gene mutations, most brain disorders present a complex interplay between the genome and the exposome through the interaction of the brain transcriptome and its regulatory network. Genetic analysis of brain disease, through profiling of tissues, cells, mitochondria, and nuclei, demonstrates common gene clusters reflected within these disorders and diseases.[199] Time will tell the precise relationship of these genes and their clusters to threat versus safety, but TVST postulates and hopes to predict their relationship to chronic threat programs within the genome.

Violence and Criminology

"It's easy to be judgmental about crime when you live in a world wealthy enough to be removed from it. But the hood taught me that everyone has different notions of right and wrong, different definitions of what constitutes crime, and what level of crime they're willing to participate in."

Trevor Noah

Threat associated with increased TL, TS and TC results in threat phenotypes that are primed for asocial, antisocial, violent, and criminal behaviors. As noted previously, threat, TS and TC within a TR impair contemplation, rationalization, organization, creation, connection, and socialization.

In threat, humans turn up the older and more primitive parts of the brain that facilitate mobilization, competition, aggression, reaction, and protection. Threat notably enhances irritability, anger, fear, sadness, despair, impulsivity, vigilance, rigidity, paranoia, prejudice, disorganization, disconnection, and asocial, if not antisocial, behaviors.

Immoral, unethical, sociopathic, psychopathic, violent, and criminal behaviors are also associated with changes in the molecular signaling, brain physiology, and connectotype of threat.[200,201,202,203] TVST postulates violent and criminal behavior are primarily phenotypes of threat and not genetic, brain, or character deficiencies or defects.

People, particularly young men, who have a chronically elevated TL, and especially those who feel disenfranchised, with limited sexual, social, and financial opportunities, are much more likely to commit criminal and violent acts.[204] Racial and ethnic minority groups are more likely to live in communities with the threats of disenfranchisement, discrimination, prejudice, indignity, inequity, injustice, and poverty. Homicide and suicide via firearms are associated with all these chronic threats.

In the United States today there are approximately 4,000 gun-related deaths per month. From 2014 to 2020, mortality from firearm homicides increased by an average of 6.9% every year.[205] The U.S. has a disproportionate level of violent deaths from extremism and terrorism, accounting for over 50% of such deaths worldwide. The cost of gun violence approaches 2% of the U.S. gross domestic product (GDP). The U.S. is a violent society and culture. A recent survey demonstrated gun culture also highly correlates with hostility, racism, misogyny, and male supremacy.

Young adults in the U.S. feel unsafe. Approximately 25% have had exposure to an active shooter situation, while 40% have access to guns, and another 17% have plans to gain access to or purchase a gun. In 2022, firearms were the leading cause of death among youth aged 1 to 19 years for the 5th straight year. Overall, there were 4,590 firearm-related deaths in that age group in 2022—the second highest rate in the last 25 years. Young adults suffer from hypervigilance (especially in public places), anxiety, depression, and PTSD at alarming rates.[206] These symptoms are not from brain or character deficiencies or defects. They are the symptoms and the phenotypes of chronic threat.

"Guns don't kill people; chronic threat kills people."

Ali F. Oeus

Strategies of further isolation, deprivation, and punishment will not only make these behaviors worse, but will cause further psychologic, neurologic, and systemic degeneration, with further declining capacities and capabilities. Accountability for maladaptive and antisocial behaviors is essential. However, increasing threat through isolation, deprivation, punishment, and dead-end incarceration only compounds these problems. The paradox of punishment is that the behavior being negatively reinforced is only more likely to recur. Norway offers another way.

Norway's incarceration rate is just 75 per 100,000 people, compared to 750 per 100,000 people in the U.S. On top of that, when criminals in Norway leave prison, they stay out. It has one of the lowest recidivism rates in the world at 20%. The U.S., by contrast, has one of the highest rates—77% of prisoners are rearrested within five years.

Norway relies on a concept called *restorative justice*, which aims to repair the harm caused by crime rather than punish people. Their criminal justice system focuses on rehabilitating prisoners. Imprisonment prepares inmates for life on the outside with vocational and social programs, not isolation and punishment.[207]

Prisons can have five goals: retribution, incapacitation, deterrence, restoration, and rehabilitation. Americans want their prisoners punished first, and at best, rehabilitated second. The U.S. Department of Justice found that strict incarceration increases offender recidivism. By contrast, facilities that incorporate cognitive-behavioral programs rooted in social learning theory are the most effective at reducing recidivism. TVST postulates that with a new paradigm, this can be improved upon.

Threat is the source of violent and criminal behavior, and safety and opportunity the cure. When people have become asocial or antisocial, they require an environment that is safe and meets their needs first and foremost to initiate rehabilitation and recovery.

Prisons must not be centers of punishment. To rehabilitate and restore a human being to sociality, prisons must incorporate an algorithm that provides physical, emotional, behavioral, social, mental, spiritual, and financial safety. They must train the incarcerated to use this algorithm as a tool to thrive when they return to society.

Terrorists are no different in their needs. Terrorism is frequently propagated by young men who feel disenfranchised, with little prospect for sexual, social, and financial success. To wage a war on them is to entrench them. Efforts to enfranchise, recruit, educate, mentor, and incorporate them would be an act of counterintelligence, as opposed to the unintelligent and unproductive strategies employed today.

Threat compounded increases violence and criminality.

Safety breaks this cycle.

Failure to follow this paradigm is simply to repeat the failures of the past and to fail to meet the needs of society and the species.

Addictionology

> *"It has not been in the pursuit of pleasure that*
> *I have periled life and reputation and reason.*
> *It has been the desperate attempt to escape from*
> *torturing memories, from a sense of insupportable*
> *loneliness and a dread of some strange doom."*
>
> *Edgar Allen Poe*

Addiction is not a disease in and of itself. Addiction is a symptom of being trapped in chronic threat.[208] Substance use is an attempt not to constrain, repress, and suppress emotions, behaviors, and thoughts, but to sedate the cortical functions that pit and conflict internalized contracts, ideations, constructs, narratives, and beliefs against emotional desires, behavior drives, and generated thoughts.[209]

Substance use is an attempt to numb the pain and suffering from the squeeze of a social or cultural trap. It is the painful pinch between the need to constrain, conform, and capitulate to belong and the desire to be emotionally, behaviorally, socially and mentally unconstrained and free. The toxic dissociation of substance use shuts off the conflicts of the mind, the noise within the human head, and soothes the psyche in the short run. However, it worsens the TL in the long run. This culminates in further dysfunction and poor health.

Alcohol, psychostimulants, THC, and opioids all (via different mechanisms) increase, not decrease, TS and TC.[210] This leaves the system catabolic, inflamed, and agitated, especially after the substance has worn off and the cortical dissociation has resolved, thus driving the desire to use and dissociate again.[211]

Additionally, TS and TC play a critical role in dopamine formation, release, breakdown, and depletion that exacerbate addictions.[212] As dopamine receptors downregulate and become insensitive, and dopamine stores are depleted, the pleasure of a dopamine high becomes more difficult to obtain over time, requiring escalating dosing to achieve the desired effects. In the end stages of addiction when dopamine is depleted and signaling is impaired, the goal is no longer to find pleasure, but simply to avoid pain and withdrawal.

Additionally, toxins induce a TR and threat-associated behaviors that can be both asocial and antisocial. Complicating this further, reactivity and impulsivity are norms in threat. Physiology drives action and reaction over examination and contemplation. This is a survival trait—better to be quick than be lunch. In addiction and in threat, behaviors can proceed awareness and cognition, whereas in safety awareness and cognition frequently proceed behavior.

It is hard to stop addictive behaviors when in threat, and it is hard to stop threat when addicted.

Addiction must be treated, along with all other threat related issues, by removing someone from internal and external threat. This is followed by discontinuing threat-stimulating toxic dissociative and addictive substances—thus dissipating maladaptive behavioral patterns. The ultimate step is to finally reintegrate into a new physical, emotional, behavioral, social, mental, and spiritual place and space of safety.

The substance is not the problem. The addiction is not the problem. The lack of safety is the problem. Trying to cure addiction through abstinence alone, aversive and punishing treatments, or worse, incarceration, will never work. Such actions will only exacerbate the illness, disease, disability, pain, suffering, and despair.

Indeed, Alcoholic Anonymous does a great job of creating

many of the requirements for being safe in the world. AA is a safe social group that allows for freedom of emotion and thought expression. AA deconstructs illusions of control, false-self protections, concepts of a disconnected self, and demands letting go of illusions of control to a "higher power." AA requires disconnecting from self-focus to serve others. Finally, AA focuses on gratitude for having basic needs met and for life itself.

It is fair to say AA helps individuals to feel safer within the world. What AA can't do is address the greater societal, cultural, institutional, and technological forces that make people insecure and unsafe in the world. AA is without the bandwidth to ensure clean air, pure water, healthy food, safe housing, financial security, societal tranquility, cultural equanimity, and institutional fairness, let alone tame marketing and media predators. To successfully treat addiction, all these issues must be addressed at a population level, as well as at an individual one.

AA also focuses on acknowledging and eliminating "character defects." TVST differentiates from AA in this area. TVST postulates that it is not genetic, brain, nor character defects that determine our emotions, behaviors, and thoughts but our total TL. Focusing on a defect in the self only aggravates the TL. To treat properly the TL must be addressed not the construct of "character defects."

Portugal has set a positive example for what can be done when drug policies prioritize health rather than criminalization. Twenty-plus years ago, facing a crisis of addiction and associated infections, the Portuguese decriminalized drugs. The results included a drop in addictions, an increase in treatment of addictions, a decline in new HIV and hepatitis infections, and a lowering of drug-associated deaths.

In addition, the criminal justice system was decompressed. The prison population fell sharply. There were enormous cost savings found in treating addiction as a health concern and not a criminal concern. Threat, illness, and disease were reduced in favor of safety, wellness, and health. Harm reduction was a priority. Naloxone provision, clean needle and syringe distribution, injection centers, overdose prevention centers, and fully funded treatment programs are central to the policy.

To get more specific, in Portugal, treatment rose 60%, drug-related deaths fell by 85%, drug-related HIV infections dropped 90%, and the overall social cost of drug use decreased by 12% in the five-year period and 18% in the ten-year period following decriminalization of drug use. The social cost of drug use is defined by the sum of public expenditure on drugs, the private costs incurred by individual drug users, as well as costs taken on by society, including loss of income and loss of productivity.

Ending the criminalization of people who use drugs is important to enable effective public health policy. However, this policy is still only one variable interacting with a complex mix of social, cultural, societal, institutional, political, and economic factors when it comes to addressing addiction.[213] Creative comprehensive changes are required to address the addiction, overdose, and death crisis of the United States.

Short of actual decriminalization, the U.S. needs to at least incorporate the harm reduction strategies of the Portuguese. But even more importantly, to lower the desire to dissociate from this world and the demand for drug use, the TL of the U.S., and the globe, needs to be reduced. To start, these changes will require addressing the indignities, inequities, and injustices—threats—within the U.S. These changes will lead to increased safety across the population.

Anesthesiology

> *"Your pain is the breaking of a shell*
> *that encloses your understanding."*
>
> *Khalil Gibran*

Anesthesiology is the study of pain and the practice of reducing pain. Historically anesthesiology has focused on the management of acute somatic more than visceral pain particularly at the time of surgery, less so post-surgery, and not until recently has there been a focus on chronic pain. To date, targeting of chronic pain has extensively used the strategies for managing acute pain. Unfortunately, this has resulted in less than desirable outcomes. Acute and chronic pain are not the same thing. TVST offers a new paradigm to conceptualize and treat chronic pain.

Over 50 million adults in the U.S. suffer from chronic pain. Chronic pain is higher among females, the elderly, the unemployed, those living in poverty, those with a disability, those in poor health with chronic medical conditions, those who are separated or divorced, those residing in nonmetropolitan areas, those with public health insurance, and veterans.[214] Those with higher TLs are disproportionately affected by chronic pain.

Generational trauma perpetuates chronic pain. Childhood trauma leads to not only depression, disease, and disability, but chronic pain and suffering in adults.[215] Minority populations can have up to two times the rates of chronic pain. The highest rates are notably in those with the comorbidities of myalgic-encephalomyelitis/chronic fatigue syndrome and dementia, two threat associated syndromes.[216]

Like addiction, pain—acute or chronic—is a symptom of threat. It is not a disease state in and of itself, it is a symptom. Threat, TS, and TC place the system on alert and activate the TR and defenses with associated systemic changes. These include inflammation and amplification of sensory inputs, including nociception. Most notably, via reactive oxygen species, arachidonic acid, leukotrienes, prostaglandins, substance P, Calcitonin Gene Related Peptide (CGRP), glutamate signaling, and ultimately increased nociceptive signaling, the pain response is amplified by threat physiology.[217]

Glutamate signaling (glutamate is the major neurotransmitter in threat states) is amplified and dominant in pain conditions. Glutamate intensifies the pain experience. Glutamate also facilitates the centralization of pain circuits and formation of predictive codes within the nervous network to create easy triggering of pain and chronicity of both pain and the TR. Some refer to these as a learned and predictive responses that can be separate from any actual tissue injury, and this can be true, but recognize that within a TR tissue without any injury can still be inflamed and catabolized. Threat physiology and pain are associated with a reduction in the production and transmission of the inhibitory neurotransmitter gamma-aminobutyric acid (GABA), and feel-good neurotransmitters such as dopamine, serotonin, melatonin, and acetylcholine. [218,219]

The periaqueductal grey matter, the primitive center of raw emotions, of the midbrain is a major relay station for the modulation of pain signaling. The PAG has a high density of receptors for monoamine neurotransmitters, peptide signalers, as well as opioids and cannabinoids. It is also a center for motivated behaviors in response to pain. Pain and associated defensive emotions, behaviors, thoughts, and autonomic functions can be modulated at the PAG well prior to the signaling reaching cortical awareness. Threat phenotypes in response to painful stimuli are coordinated at this level well below cortical awareness. Yet, the cortex also feeds back to this level. The phylogenetically newer medial prefrontal cortex and the somewhat older anterior cingulate cortex are centers where emotional and social injury, pain, and conflict give input to the more primitive PAG to influence its functions. Distress and pain signaling from both above and below can amplify a TR and amplify the sociosomatoemotiovisceral experience of pain.

When in safety the PAG expresses the phenotypes of safety with relaxation, and social, sexual, and maternal behaviors where serotonin, oxytocin, and parasympathetic tone are high, and pain in less.

Acute pain can very much reflect the T1 phenotypes, and chronic pain tends to reflect the T2 phenotypes. These states are phenotypically and physiologically very different. Chronic pain has less to do with physical injuries and external and peripheral inputs and inflammation, and more to do with social and emotional injuries with internal and central inputs and less inflammation. However, both involve persistent catabolism.

"The only antidote to mental suffering
is physical pain."

Karl Marx

Chronic pain syndromes reflect the cumulative, compounded, and chronic TL. They are tightly associated with the brain centers for interoception, visceral and fascial sensations, and emotional, behavioral, social, and mental conflict and pain. Most notably the prefrontal, cingulate and insular cortices, and the nucleus accumbens are involved with chronic pain conditions.[220,221,222,223]

Complex regional pain syndromes (CRPS) are threat-related syndromes with increased TS, TC, and an amplified and sustained TR. CRPSs have two phenotypes. One is a "warm" phenotype that appears much like a sustained acute pain response and is consistent with a T1 phenotype. The other is a "cold" phenotype, with less inflammation, persistent catabolism, hypometabolism, wasting, loss of bone and cartilage, fibrosis, and contractures that are all consistent with a T2 phenotype.[224,225]

Treating chronic pain syndromes with structural and acute pain strategies is infrequently helpful and sometimes harmful. Alternatively, treating chronic pain with emotional, behavioral, social, mental, and spiritual interventions to extinguish maladaptive illusion of control and decrease repression and suppression is remarkably helpful. This all helps reduce threats. Treatment methods using somatic and emotional integration, free expression, and safe social connection work well in the treatment of chronic pain conditions. Changing false created ideations, cognitive constructs, symbolic narratives, and held beliefs also helps in managing chronic pain. In fact, these methods work not only for chronic pain conditions, but for all nonphysical threat-related conditions. [226,227,228]

It is important to be able to recognize these threat phenotypes and physiologies, as well as to unroot the total TL, when assessing pain and pain syndromes. But perhaps the most important factor is to simply recognize the root of pain is threat and work to move to safety. This includes assessing an individual's total TL and then systematically resecting threats and infusing safety.

Physical, emotional, behavioral, social, mental, cultural, societal, financial, and spiritual threats must be included in a comprehensive evaluation and treatment plan for disease, disability, pain, and suffering.

"Because there is no love you can throw on them, no hug big enough that will change the power of that drug; it is just beyond imagination how controlling and destructive it is."

Beth Macy

It needs to be noted that opioids, frequently used for pain management and the center of the current pandemic level of drug abuse, addiction, overdose, and death, are endogenous (endorphins, enkephalins, dynorphins) signalers within a threat response. Opioids turn off the reproductive and gastrointestinal networks. Opioids down regulate sex hormones and bonding. Social disconnection, asociality, if not antisociality, are increased with exogenous opioid use and addiction. Opioids are proinflammatory and catabolic. Opioids induce wasting and immuno-suppression when used chronically. Opioids amplify glutamate signaling, resulting in engrained and intensified pain pathways. Opioids amplify TL, TS, and TC.

A recent study on the use of opioids for chronic neck and back pain demonstrated opioids were no better than placebo, and arguably they are much worse.[229]. Given all of this, there are simply no indicators to support the chronic use of opioids. Opioids should only be used short term in acute pain conditions and during active dying.

Americans consume 80% of the world's opioids, despite accounting for only about 5% of the world's population. There are multiple reasons for this phenomenon, but Americans do not have different sensors or signalers for pain to account for this situation. Under the lens of TVST, the most striking issue accounting for this phenomenon would be Americans' TL leading to increased levels of disease, disability, pain and suffering, and associated opioid use to dissociate from the pain of life. Instead of dissociating from pain and suffering, it would be better to treat the source of the pain and suffering, treat the root—chronic threat.

This argues that those who live in isolation, poverty, discrimination, disenfranchisement, indignity, inequity, and injustice would be more likely to self-treat or be treated with opioids—and they do and are. Negative social determinants of health predict the risk for opioid prescription, opioid use disorder, and opioid associated deaths.

A second issue may be Americans' expectations to lead a pain-free, if not pleasurable life. This is not realistic nor possible any more than leading a threat free life, or not aging, is possible. Utopia or heaven is not achievable on Earth. Threat and pain are part of life, and efforts to eliminate pain completely are folly. Pain is essential for life, and the only way to completely eliminate threat and pain is to die. Pain is a sentinel for threat, to be respected and regarded. Pain can be minimized, a worthy goal. However, the target should not solely be the reduction of pain, but the reduction of suffering—chronically being trapped in threat. Chronic opioids have no role in this endeavor as they amplify both pain signaling and suffering.

What looks like America's opioid crisis is not a crisis of opioid use, but a crisis of suffering from being trapped in chronic threat. Disease, disability, pain, and opioid use are the symptoms, not the source, of this crisis.

Pain is a normal signal for danger, and only sometimes a signal of damage. When trapped in threat, chronic pain is a normal biophysiological response—danger is near. Pain is a normal physiologic response to threat, but maladaptive in chronicity. Existence within chronic threat is not consistent with wellness, health, and happiness.

Chronic pain is not necessarily a sign of being broken. The body is doing exactly what it is supposed to do when danger is perceived to be near. In no way are opioids a cure for this state of being, as they deepen threat signalings. Relief from chronic pain through masking, numbing, and dissociation without addressing the source is to neutralize a danger signal without removing the danger. If signals were removed from highways, disaster would follow—see the opioid crisis. Opioids do not treat or cure the sources of pain and suffering but create an illusion of control. This, paradoxically, makes humans more reckless and unsafe.

"I have a big fear of things spiraling out of control.
Out of control and dangerous and reckless and
thoughtless scares me because people get hurt."

Taylor Swift

To reiterate, clinicians are only able to affect 20% of the determinants of health. The other 80% fall grossly into the behavioral and social determinants of health. Therefore, clinicians are only reasonably able to affect 20% of the determinants of chronic pain. The treatment for the other 80% lies elsewhere within the total TL.

Individual responsibility and behavioral modification to lower the TL are important in pain management (and a challenge as threat physiology dictates behavior). However, perhaps more important to holding the individual accountable to their behaviors is the lowering of the societal and cultural TL. Through good societal policies, politicians may be able to do far more than clinicians in terms of chronic pain "management" and to resolve our pain, suffering, and opioid crisis.

The poor societal safety net in the U.S. requires people to medicalize their status to obtain any help. Amplification of disease, disability, pain, and suffering are a consequence of missing or failed societal policies, not individual character flaws, deficiencies, or disorders.

Pain is not a disease, but a modality by which humans sense threats. It is a symptom of threat, inclusive of physical, emotional, behavioral, social, mental, and spiritual threats. Pain is essential for survival. It is time to evolve the concepts and treatment of chronic pain and suffering.

Placebo

> *"We often give painkillers the credit that ought to be given to time, the belief that they would kill the pain, or the water that accompanied them."*
>
> *Mokokoma Mokhonoana*

A placebo can be roughly defined as a sham medical treatment. Common placebos include inert tablets (like sugar pills), inert injections (like saline), and sham procedures. Placebos are used in randomized clinical trials to balance out the effect of simply getting a treatment against the specificity of the drug or procedure being studied.

Placebos are felt to sometimes produce relief through psychological mechanisms. This phenomenon is known as the "placebo effect." Placebos are felt to change how patients perceive their condition and thus relieve pain and/or other symptoms. However, within this context they are felt to have no impact on the disease itself. The placebo effect is usually short-lived, but the effect can last thirty days or longer.

Placebo pills and procedures are also unknowingly used every day in clinical care with a belief that the pill or procedure worked or will work, a belief the intervention is not a placebo by both the patient and the provider. The placebo effect gives the impression the intervention is working, but ultimately the effect is short lived. Confirmation bias and belief result in the interventions being repeated over and over again, even though they are not effective in the long term. Unfortunately, this can be a costly practice, both financially and in terms of harm done.

TVST does not view placebos as sham treatment or the "placebo effect" to be a purely psychological mechanism. TVST purposes that the "placebo effect" is simply the provision of a safety ideation, construct, narrative, and belief that temporarily shifts the phenotype and physiology into a SR. This has a physiological and, yes, thus a psychological effect, and an effect on the disease. However, it is typically temporary, and not a cure. The total TL remains untreated and when the SR wears off, the symptoms return.

The practice of using temporary safety ideations, constructs, narratives, and beliefs—placebo effects—is common within the practice of chronic pain medicine, and particularly in spine medicine. Most neck and back pain is myofascial and physiological, and not musculoskeletal and structural, in its etiology. Minor procedures and the "placebo effect" are used to justify more dangerous and more costly procedures. The rationale is that a structural problem has been identified when patients declare improvement from minor structural interventions. When only the "placebo effect" is at play, patients can be harmed and suffer more in these scenarios—and these scenarios are not rare.

Medicine should use the "placebo effect" for legitimate, not deceptive purposes (in some cases, both the patient and provider are deceived). There is never a contraindication to stimulating a SR. However, sustaining this response is frequently a much heavier lift than perceived. The total TL needs to be addressed. Success comes when both a SR is induced and a TR is reduced, and safety is sustained.

Neurology

> *"I think we're all fascinated by how the brain works. One of the most mysterious of the physical sciences is neurologic science."*
>
> *Alexis Denisof*

Brain disorders, including mental illness, neurologic conditions, and stroke, account for a significant illness and financial burden globally. Combined, they account for more than either cardiac disease or cancer per the Global Burden of Disease (GBD) study.[230] The trends show:

- By 2050, more than 50 million people will be aged 65 to 79.
- The number of individuals aged ≥ 65 years will increase by 350% by 2100.
- Alzheimer's will increase by 226% between 2015 and 2040.

- Depression is estimated to affect over 300 million people across the globe, which represents a 71% increase since 1990.
- The numbers for anxiety-related disorders have similar trends.
- Despite advancing technologies, pharmaceuticals, and procedures the number of strokes has increased by 95% since 1990.

To estimate the toll caused by brain conditions, including brain injuries, mental disorders, neurologic disorders, cerebrovascular disease, brain cancer, and select infectious conditions, researchers have calculated the morbidity and mortality associated with brain conditions. Neurologic afflictions of some form have been estimated to affect 3.4 billion people, which is approximately 43% of the world's population. This has resulted in 443 million years of healthy life lost, and 11 million deaths annually. This is up 20% from two decades ago. Globally, brain conditions accounted for more than 15% of all health loss in 2021, slightly more than cardiovascular disease and almost double cancer's health loss.

This health loss is associated with a $1.22 trillion loss in income for people living with health disorders worldwide and accounts for $1.14 trillion in direct healthcare costs. The burden of mental disorders, neurologic conditions, and stroke is expected to increase dramatically between now and 2050. Not all of this can be accounted for within the statistics of an aging population. A very worrisome trend is the rising health loss linked to brain conditions in younger people.[231] A rising global TL can account for these findings.

Many factors correlate to brain disorders, including the typically noted things such as education level, obesity, and smoking. Therefore, it is felt that a healthy brain can be achieved through an education, healthy lifestyle, and the management of conditions such as high blood pressure and diabetes, eliminating alcohol consumption and smoking, prioritizing sleep, eating healthy, and staying physically and mentally active.

And all of that is true. But if one steps back from the elephant, it becomes apparent that TVST explains these trends and would serve as a good platform from which to correct course.

It should also be noted that some neurodegenerative syndromes follow a very similar pattern and trajectory to the psychological and cognitive disorders of threat. Those who suffer from depression have 2.5 times the risk of suffering from dementia.[232] In addition, the symptoms of depression mirror the symptoms of dementia. However, TVST postulates that this correlation should not be confined by diagnostic rigidities such as depression and dementia, as both are on the continuum of a chronic TR. Depression and dementia follow similar physiologic patterns and processes without one necessarily being a risk factor for the other, as they both arise from the same root and soup of threat. They are correlative, but not causative.

Similar prioritization, resource allocation, and systemic dissolution patterns play out over a longer period of time in neurodegenerative disorders versus mental health disorders. This slow and more subtle process is a gradual downregulation, deconstruction, destruction, and decline, a stealthy neurodegeneration, that results in the delay in diagnoses and treatment.[233,234,235,236] These disorders can start as T1 phenotypes then progress to T2 phenotypes as aging advances and the TL is sustained, increased, or compounded.[237]

The classic example of a T1 phenotypic neurodegenerative disease is Alzheimer's-type dementia (AD), with primary symptoms of partial neocortical dissolution, dysfunction, and degeneration. This relationship to threat, along with TS and TC, is being more clearly established.[238,239,240,241,242] Distress affects brain structure and inflammatory processes, with chronic distress having increasing detrimental effects on brain structures especially the human neocortical structures. Evidence suggests a role of chronic distress in the pathogenesis of AD. Accumulated stressful life events correlate with AD pathologies such as neuroinflammation and reduced brain volumes among cognitively unimpaired individuals at heightened risk of AD. A recent study demonstrated that stressful life events were associated with AD pathophysiology, or perhaps better said AD phenophysiology, and neuroinflammation. Among those with a history of psychiatric disease stressful life events were associated with higher tau protein and TC. Participants with a history of psychiatric disease showed reduced brain volumes in neocortical, somatic, and limbic regions.

In AD, declarative memory and symbolic language impairment are frequently the first symptoms to be noticed, but social and intellectual dysfunction are also a part of early AD. Mobility can be well preserved until the later stages of the disease.

The classic example of a T2 phenotypic neurodegenerative disease is Parkinson's disease, which has deeper and phylogenetically older brain structures undergoing dissolution, dysfunction, deconstruction, destruction, and degeneration in a chronic immobilization response. The relationship between Parkinson's disease and threat, TS, and TC, is well established.[243,244,245]

Indeed, chronically elevated TS, TC, and TLs have been established in numerous neurodegenerative disorders.[246]. Autoinflammatory, auto-immune, cardiovascular, and gastrointestinal diseases have also been associated with "increased risk" for Parkinson's disease.[247] In fact, inflammatory bowel disease and Parkinson's disease share genetic, epigenetic, and metabolic profiles—same soup presenting in different cells, tissues, and organs. These different diseases and symptoms are more correlative than causative; they all simply arise from the same root and soup of threat.

Nearly all chronic physical, emotional, behavioral, social, mental, and spiritual health issues arise from the same root of chronic threat.

The Lancet Commission in 2017 concluded there were nine major risk factors for Alzheimer's dementia:

- Less education
- Hypertension
- Hearing impairment
- Smoking
- Obesity
- Depression
- Physical inactivity
- Diabetes
- Low social contact

In 2020 three more risk factors for dementia were added:

- Alcohol consumption
- Traumatic brain injury
- Air pollution

Recently, separate from income and education, simply living in disadvantaged neighborhoods was added to this list of risk factors. Living in these neighbors was a risk for grey (cortical) matter shrinking, white (connections) matter abnormalities, premature brain aging, impaired cognition, and dementia.

TVST suggests these risk factors are a smattering of threats, true risk factors, threat phenotypic changes, and correlates. Threats include a stressful environment, less education, any pollution, smoking, alcohol, injury, and isolation. These are causal. Threat-associated phenotypic changes include hearing impairment, hypertension, obesity, diabetes, and depression. These are phenotypic threat correlates more than risk factors. Nevertheless, total TL is at the root of dementia.

A recent study demonstrated a person's proteome (peptides and proteins) has markers or correlates for dementia that are present years before onset of apparent symptoms. In addition, not all the proteins showed changes in both plasma and brain tissues. This suggests systemic correlates that are not specific to the brain itself. Many of the correlating protein molecules are consistent with an inflammatory and immunologic defense response, and others more consistent with a metabolic response. Yet all appear consistent with the TVST model of Alzheimer's disease being a threat-related disorder.[248]

In Alzheimer's disease, the nonneuronal brain cells, microglial cells, assume proinflammatory and catabolic threat phenotypes. If threat is perceived, then why wouldn't this be a systemic phenotype? Well, peripheral white blood cells also assume threat phenotypes that correlate with Alzheimer's disease, thus demonstrating the systemic nature of the disease. To only see the brain's involvement is akin to the parable of the Blind Men and the Elephant. It is to miss the whole while studying the parts, to miss the truth. Perhaps better said, the cause of the dementia isn't extracellular beta amyloid plaques, intracellular tau protein tangles, multiple APOE4 genes, or even something named "Alzheimer's Disease"—it's chronic, cumulative, and compounded threat.

Worldwide, around 50 million people live with dementia. This number is projected to increase to over 150 million over the next 30 years, rising particularly in the less fortunate. Dementia affects individuals, their families, and the economy, with global costs estimated at about $1 trillion annually. [249] The solution for this social and economic burden of a systemic process with the symptom of neurodegenerative is unlikely to be found with a single targeted pharmacological intervention alone. Creating a safer world on all fronts will lessen this burden.

"Schizophrenia is one of the most misunderstood diseases on earth. It is a physical brain disease, like Alzheimer's, Parkinson's and stroke, but more treatable."

Bethany Yeiser

It is also important to recognize that schizophrenia is a neurodegenerative disorder and a threat-related disorder that crosses multiple medical silos encompassing psychological and neurological changes, as well as a vast array of other comorbidities. TVST demonstrates the mechanisms for the behavioral, neuropsychological, and neurodegenerative changes of schizophrenia.

Schizophrenia is a severe threat spectrum dissociative disorder that culminates in degenerative neurologic changes. The use of addictive toxins to sooth the soul will accelerate the disease. Loss of financial and social supports will do the same. Schizophrenia has both T1 and T2 phenotypes. T1 phenotypes are characterized by:

- Irritability
- Impulsivity
- Grandiosity
- Verbosity
- Hyperkinesis
- Rejection

By contrast, T2 phenotypes are characterized by:

- Isolation
- Pessimism
- Rumination
- Obsession
- Dyskinesis
- Hopelessness
- Helplessness

Observable phenotypic postures and facial expressions are also associated with these states to go along with stereotypic behaviors and the negative, intrusive, paranoid, automatic thoughts.

Whether in a T1 or T2 state, what is most striking in schizophrenia is attenuation of the cognitive control and salience networks, enhanced connectivity within the default mode network, and profound neocortical dissociation with emerging delusions and hallucinations likely bubbling to awareness from the codes running below the neocortex. Arguably, treatment should focus on threat and safety and decoding and recoding—no small tasks—more than pharmacologic management of symptoms. Treat at the root.

Schizophrenia is a profound piece to the proof that humans fundamentally do not control their thoughts and that their thoughts may not reflect reality. Emotions, behaviors, and thoughts are a better reflection of the weight of the TL than the state of character, aspirations, desires, and will. Reducing the total TL is more important than changing perceptions, unless perceived threat is the bulk of the TL. Threat-related perceptions within a human being suffering from schizophrenia are typically the symptom of the threatened soul and not the bulk of the TL. This notion is essential to the paradigm of TVST and the development of an algorithm for healing, health, and happiness.

> *"What do you think you are, for Chrissake, crazy or somethin'? ..... Well you're not! You're not! You're no crazier than the average asshole out walkin' around on the streets and that's it."*
>
> *McMurphy – One Flew Over the Cuckoo's Nest*

Ruminations and aversive, intrusive, and obsessive thoughts are warnings that TL is high. Delusions and hallucinations are confirmations of an unheeded warning. These aversive thoughts are not to be ignored. However, spinning them into toxic ideations, flawed constructs, false narratives, and rigid beliefs is maladaptive. Awareness of how humans operate and proper integration of this physiologic phenomenon of thought into daily life will be crucial to a successful treatment algorithm. Treat the root, change the physiology, and better thoughts will follow.

Safety is the treatment of choice for all neurodegenerative disorders, yet very different strategies are used in the care of someone suffering from schizophrenia versus Alzheimer's or Parkinson's disease. The U.S. doesn't do well with any of these diagnoses. However, the neglect of a patient suffering with schizophrenia within U.S. society and culture is striking and exemplified by how many people with schizophrenic symptoms live on the streets—something unacceptable for victims of other neurodegenerative diseases.

"The homeless often feel invisible, allowed to plummet through the widening holes in the social safety net, then hidden in doorways from which people avert their eyes."

Dawn Foster

Over 10 million Americans suffer with schizophrenia, and over 20% of the approximately 700,000 homeless Americans suffer with schizophrenia. Homelessness is the epitome of not being safe, seen, nor secure—threat. Minority populations are disproportionately distressed, homeless, and at risk for neurodegeneration. The suffering seen with schizophrenia is not accepted in other neurodegenerative disorders such as AD and PD. Alzheimer's is seen as a neurological disease, grounded in an excess of beta amyloid. By contrast, schizophrenia is seen as a psychological disease, characterized by a loss of reality. Modern Western culture, which associates mental illness with weakness in character and mind, as opposed to threat-associated faltering physiology, is in fact responsible for such mental illness. Marginalizing and dehumanizing follows these cultural ideations, constructs, narratives, and beliefs, and allows for not caring.

In addition, schizophrenia appears in relatively youthful persons, whereas Alzheimer's appears in relatively older ones. The young are supposed to be capable and able to provide for the old.

These two syndromes are not equated despite the similar physiology and brain changes associated with them. One group needs to pick themselves up by their bootstraps; the other is to be pitied. Both diagnoses follow the pattern of chronic threat physiology and deserve treatment for threat.

The Swedish have a different model for psychologic healing. In this era of multi-drug cocktails and psychiatric multi-diagnoses that stick with patients for a lifetime, the Swedish model helps people recover from psychosis without medication or attention to diagnostic constructs. Counseling, institutionalization, and pharmacologic interventions have failed as the pillars of care for psychiatric patients. Frustrated by wasted resources, failed policies, misaligned treatments, false constructs, and poor outcomes, Sweden pivoted. Instead of consigning the solution to the individual, Sweden now takes a more holistic and comprehensive approach to mental illness. This approach includes addressing poverty, education, jobs, rights, and social connection at a societal level.

Backed by over twenty years of experience, Sweden now places people who have been failed by traditional medicine and psychiatry into host families, predominately farm families in the Swedish countryside. Host families are chosen not for any psychiatric expertise. Rather, they are selected for their compassion, stability, and desire to give back. Patients live with these families for upwards of a year or two and become an integral part of a functioning family system. Social and mental health services are still provided, including psychotherapy, and providing host families with supervision and support. This program eschews diagnoses and works to wean medications. All this is provided for free within the Swedish social and health service.

It would be remiss not to mention that this model of care has roots as far back as the 13th century in Geel, Belgium. Oliver Sacks, M.D., a famous neurologist and author, who died in 2015, wrote about the alternative Geel paradigm for mental health care:

> *Is there a chance that a more human and social approach, trying to reintegrate them into family and community life, a life of love and [labor], will succeed as well? Even those who could seem to be incurably afflicted can, potentially, live full, dignified, loved and secure lives.*

TVST provides the biology behind this phenomenon and highlights why traditional counseling and pharmacologic intervention fare so poorly by comparison.

The popular monoamine—largely dopamine, but also serotonin—theory of schizophrenia is felt to be misguided. While blocking dopamine transmission helps with the threat-based defensive behaviors of schizophrenia, it fundamentally impairs emotional, behavioral, social, mental, and spiritual processes and integration. In turn, this blocks healing and recovery.

"First, do no harm."

Hippocrates

TVST defines beta amyloid, tau protein, and alpha synuclein not as toxins within the brain, but as metabolic (catabolic) hyperphosphorylated dormant protein artifacts associated with threat, TS, TC and a TR that includes cellular downregulation, senescence, catabolism, and deconstruction. These changes are seen within threat phenotypes. Beta amyloid, tau protein, and alpha synuclein have not only contextual roles to threat, but contextual roles to safety where they are responsible for cortical upregulation, neuroplasticity, anabolism, and construction.

The changes seen in neurodegenerative diseases are a result of chronic threat programing and aging physiology.[250] TVST suggests that removing these artifacts will not have a significant impact on these diseases—they are not causative, just correlative. Case in point, monoclonal antibodies targeting these proteins have not been very successful in managing or reversing neurodegeneration. Adverse side effects have been of great concern with these medications, as well. Removing these artifacts may be harmful. These proteins have significant roles in cellular regeneration, as well as healthy and needed cellular deconstruction, recycling, and repurposing. Beta amyloid also has a significant role in immune and metabolic functions, specifically lipid mobilization.[251]

Normal deconstruction, destruction, recycling, and rejuvenation processes of aged and dysfunctional biologic structures are impaired in hypoactive T2 phenotypes and are reflected in hypoactive brain glial cell function—the cells responsible for these functions. This may result in less recycling and more waste and residue accumulation within the brain, with associated further impairments in brain functions. Nevertheless, this shouldn't be misconstrued to mean amyloid, tau, and synuclein are toxins, per se. A threat-associated hypoactive waste mobilization network is the more likely culprit for impairment within this stasis phenotype.

Monoclonal antibodies directed at the drivers of a TR at a systemic level could be of more benefit than targeting beta amyloid within the brain. But safety signaling may be even more beneficial. Safety restores these cells and cellular functions.

Slowly, neuroscience is moving away from an amyloid theory for the etiology of Alzheimer's disease. Other theories purposed have been mitochondrial dysfunction, glial dysfunction, autoimmune disease, gut dysbiosis, chronic infections, and metal metabolism impairment. Step back to see the cascade of physiology that is an elephant of chronic threat: AD is a systemic auto-immuno-inflammo-metabolic disease of chronic threat.

"Leaders win through logistics. Vision, sure. Strategy, yes. But when you go to war, you need to have both toilet paper and bullets at the right place at the right time. In other words you must win through superior logistics."

Tom Peters

Recently serum glycan levels have been correlated with a risk for Alzheimer's disease. Glycans are sugars that are mobilized within threat states for the fuel required to mount a defense. This new finding should not be confused with causation, but noted as a correlate, and perhaps, confirmation of systemic threat physiology being central to neurodegeneration.[252] When thinking of neurodegenerative diseases such as Alzheimer's and schizophrenia, we must consider threat phenotypes and how useless the functions of the "sapiocortex" are in a fight with a pathogen or predator. The newer "sapiocortex" is thus vulnerable to dissociation, degeneration, and dissolution—last in, first out—in cumulative, compounded, and chronic threat states.

Those living with schizophrenia and Alzheimer's fare much better and require less medication when provided with the fundamentals of safety—clean air, pure water, healthy food, access to healthcare, secure housing, expressive freedom, close social connections, positive cognitive constructs, and reintegration with the natural world.

TVST argues that with slow, degenerative processes, treating all the aspects of a TL early—prevention—will be easier, more efficient, and more effective than trying to reverse and cure the disease late. When the TL has accumulated and is compounding, it is difficult to reduce. Policy and societal resource prioritization and allocation should reflect this strategy.

Pharmacologic interventions in neurodegenerative diseases should be focused less on neurotransmitter modulation or the removal of metabolic artifacts, and more on the fundamentals of TS inhibition and SS amplification.[253]

One final note to this section is that the proper treatment of concussion, mild-severe traumatic brain injury, and chronic traumatic encephalopathy will be found within the TVST paradigm and algorithm. A concussion is most consistent with the glial cells sensing and signaling threat—a physiologic response, not a structural injury. Second, impact syndrome is the equivalent of a focal cytokine storm within the brain. Chronic glial threat phenotypes have been associated with Alzheimer's disease, and predictably the same phenotypes within a chronic TR can account for the symptoms and morphology of CTE.

Endocrinology

TVST provides a platform to explain metabolic syndrome and all the downstream symptoms and comorbidities of this syndrome.[254] Threat, TS, TC, and the associated TR induce metabolic changes, strategically alter fat metabolism and storage, shift lipid formation to increased inflammatory low-density lipoprotein-cholesterol (LDL-C) and triglycerides, and increase insulin resistance, resulting in hyperglycemia and Type 2 diabetes mellitus.

Threat, TS, TC, and the associated TR also induce catabolism with degenerative changes, such as osteoporosis, sarcopenia, degenerative arthritis, and neurodegeneration.[255]

"Shrinking someone's stomach to the size of a walnut with surgery doesn't address the underlying causes of obesity."

Mark Hyman

Obesity

Obesity with sarcopenia (OS) is particularly worrisome for a deep threat state. OS is associated with a substantial increase in all-cause mortality. Obesity can occur concurrently with a malnourished and catabolic state if in T2 physiology. The obese are not necessarily over nourished. Chronic threat, TS, and TCs induce not just inflammation but also catabolism where cachexia and muscle wasting can be seen.

Counting and restricting calories and increasing exercise does little to address chronic threat and perhaps worsens chronic threat. Chronic threat, TS, TC, and the associated TR, are amplified in obesity. Obesity sustains and compounds inflammation and catabolism via white adipocytes (fat cells within adipose tissue) releasing additional TC, adipokines, to further aggravate the system. White adipose tissue deposition is highly correlated with chronic disease, disability, pain, and suffering (obesity, T2DM, metabolic syndrome, obstructive sleep apnea, gout, heart disease, kidney disease, liver disease, pancreatic disease, arthritis, cancer, dementia, depression…), along with past traumas.

Obesity has been correlated to higher levels of distress and elevated cortisol levels. However, in fact, it is more common in a T2 phenotype to see lower and dysregulated cortisol despite worsening obesity. By stepping away from the hypothalamic-pituitary-adrenal hormonal network to the peptide network for a moment, one can find a better explanation for this phenomenon. The cytokines determine inflammation, oxidation, immunity, metabolism, lipid changes, and fat phenotypes and distribution. They declare systemic physiology from threat to safety.

White adipose accumulation around the midline and viscera is both immunologically and mechanically protective of the ventral vital organs when under attack from either a pathogen or a predator. Accumulation of white inflammatory fat within its classic apple distribution is a coded response to a chronically elevated TL.

In contrast, safety signaling via beta 3 adrenergic, acetylcholine, and GABA transmission increases brown fat, reduces white fat accumulation, and improves lean body mass. Brown fat is anti-inflammatory, hypermetabolic, thermogenic, and accumulates around the great vessels in the dorsal abdominal cavity. The toxic white fat phenotype and associated metabolic syndrome (obesity, hypertension, dyslipidemia, hyperglycemia, insulin resistance, and type 2 diabetes mellitus) have at their roots chronic threat. Look close to see this is more a threat phenotype than a disease, disorder, or a syndrome.

The threat phenotypes cause emotional, motivational, behavioral, social, and mental changes. Threat, TS, and TC, as well as other peptide signalers and hormones, change the drive for food and the foods that are craved. "Food addiction" may have less to do with dopamine and pleasure—"hedonic" eating, or serotonin, acetylcholine, and vagal tone—"emotional" eating—but more to do with threat and survival—the drive to eat foods that are easily digested and stored as a future energy source for predicted hard times.

Regardless, whether considered hedonic, emotional, or threat-based, the drive to eat is always physiologic. Emotions are a representation of physiology. There is no physiology-psychology separation. There is no mind-body duality. To divide the system in this fashion is a false construct that can lead to false paradigms and poor treatment strategies. TVST deals with the physiology of chronic threat and obesity primarily and the associated correlates to this physiology secondarily. Chronic threat is the primary source of toxic obesity, the emotions, motivations, or behaviors, arise secondarily.

Obesity, type 2 diabetes mellitus, and metabolic syndrome are threat-related disorders. The global burden of obesity and diabetes is rapidly approaching a billion people. A study on the global burden of diabetes concluded that there are many complex social and economic dynamics at play when it comes to trying to live a healthy lifestyle (behavioral and structural factors related to maintaining a healthy diet and getting enough physical activity), and that low income, low education, and living in urban areas are all associated with a higher risk of developing type 2 diabetes. Additionally, type 2 diabetes disproportionately affects indigenous populations across the world, largely attributed to colonization and disruptions to their traditional ways of life.[256]

Obesity is generational. Genetic and epigenetic coding for obesity is transferred to offspring. Having two obese parents increases the likelihood of being obese as an adult sixfold. Even living in a traditionally "redlined" neighborhood—racially segregated areas where banks and mortgage companies refused to provide services—is associated with increased risk for obesity and diabetes.

Notably, metabolic and inflammatory syndromes are increased in adults with a history of adverse childhood events, in lower socioeconomic status populations, and in the disenfranchised and discriminated.[257] Poverty is the highest risk factor for these maladies, but lack of education, substance use, unemployment, incarceration, and living in food deserts with lack of access to healthy nutrition are all risk factors. They are all threats.

Is it proper to conclude "trying to live a healthy lifestyle," "living in urban areas," or "disruptions to their traditional ways of life," are at the roots of this problem? Or is more precise to conclude chronic threat is at the root, and threat physiology is expressed in phenotypes consistent with inflammatory and metabolic maladaptation? TVST proposes the latter.

Additionally, obesity is now seen as a leading cause of cancer. There seems to be confusion of correlation and causation. Obesity correlates with cancer, as they both arise from the same root and soup, chronic threat.[258] Obesity and white adipocytes increase TC and TS. So yes, obesity influences the system and accelerates the phenotype of threat, including pulmonary disease, heart disease, kidney disease, liver disease, neurologic disease, psychologic disease....and cancer.

The T2 phenotype is associated with fatigue, a lower metabolic rate, repressed energy expenditure, hypothyroidism(itis), and hypoadrenalism—conservation for preservation—to protect against excessive combustion and degeneration. Fundamentally, this is an ancient deep threat phenotype designed to sustain the genetic code, gone awry within the modern world of human beings.

Of note, mitochondria sensing threat undergo fission. They stop oxidative phosphorylation and antioxidant formation in favor of reactive oxygen species, oxidative enzyme production, inflammation, and decreasing fat metabolism. This makes cells more dependent on sugar as a fuel for energy production. In threat, humans do not burn (beta oxidation) fat for energy very well. In threat, fats are used for other purposes in protecting the code. This reduction in beta oxidation associated with mitochondrial threat phenotypes is directly associated fat storage and weight gain.[259]

Threat, TS, TC, and the associated TR also bias the system to choose high energy easy to digest foods, i.e., simple carbohydrates. This makes sense when running from the tiger. Sugars are easy to digest and thus the preferred food. The gut is relatively offline for the digestion of complex carbohydrates, proteins, and fats when in threat. What appears as "bad food choices" are completely consistent with a threat phenotype.

Additionally, behavior can come before thought, awareness, and choice particularly when in threat states—the choice is made in advance of cortical awareness.

In threat states, if food is readily available cheap, satisfying, sugary, then simple carbohydrates are physiologically preferred. The combination of fatigue, low metabolic rate, simple carbohydrate intake, insulin spikes, increased fat storage, altered fat metabolism, insulin resistance, escalating inflammation and oxidative stress, and progressive obesity simply lead to worsening metabolic syndrome and further elevation of TS and TC. This combination results in the acceleration of threat-related comorbidities, dyslipidemia, hypertension, cardiovascular diseases, fatty liver disease, kidney disease, neurodegenerative diseases, and cancers.[260]

Breaking this cycle is crucial to the resolution of metabolic syndrome and the restoration of metabolic health. Safety, SS, SC, and an associated SR are keys to unlocking this trap.

"Let food be thy medicine and
let medicine be thy food."

Hippocrates

It is worth noting here that if glucose is the preferred or selected fuel for T1 phenotypes—fortress, fight, flight, and fever, and it is, fructose is the preferred or selected fuel for T2 phenotypes—freeze, falter, and faint. Excess glucose will be stored as glycogen and fat, but glucose fuels a high metabolic rate. By contrast, fructose slows metabolism, increases foraging behaviors and food intake, stimulates distress hormone production, and heightens insulin resistance, in addition to leptin resistance. It also induces water retention, glycogen storage, and fat accumulation. This all serves to protect and preserve the organism during hard times or predicted hard times. In addition, fructose increases insulin resistance in and decreases blood flow to the high-energy consuming neocortical brain structures, causing cognitive dulling, if not degeneration.[261,262]

From an evolutionary perspective, high fructose intake occurred in the late summer and early fall as foods ripened—a synchronicity in anticipation of limited resources throughout the winter as part of a survival program. This was a coordinated cycle that was not sustained year-round and followed by periods of prolonged caloric deprivation and ketosis. The modern world has constant fructose sources available year-round and constant complex threat at every turn. T2 physiology supports fructose intake, and fructose intake supports T2 physiology. Threat and fructose compound each other. This leads to (among others):

- Inflammation
- Oxidative stress
- Water retention
- Hypertension
- Obesity
- Insulin resistance
- Type 2 diabetes
- Metabolic syndrome
- Cardiovascular disease
- Thromboembolic disease
- Liver disease
- Kidney disease
- Neurodegenerative disease
- Cancer

Interestingly, ketone bodies can reverse this pattern. Overnight fasting allows for transient ketosis that can advance metabolic health and safety physiology by turning off inflammatory and catabolic processes. However, extended fasting and starvation diets should be avoided, as they can lead to hypometabolic changes and additional harm.

Starvation dieting for weight loss has clearly been demonstrated to be without sustained success. Only 20% of people can sustain their weight loss, and for some, the rebound weight gain exceeds their pre-diet weight.[263] Severe caloric restriction is sensed by the system as a lack of resources, a threat, and thus, biases coding and physiology towards T2 phenotypes with fatigue, immobilization, hypometabolism, defensive activation, and fat accumulation for predicted hard times. In addition to fat loss, muscle mass is depleted when malnourished. This too has implications for further lowering basal metabolic rate, activity, and anabolism.

Food restriction changes the brain to be more reinforced by the consumption of food, but also by risky and addictive behaviors. Dopamine is depleted in starvation, yet sensitivity to dopamine is increased. This makes some sense. When resources are scarce, then food should have more relative value and be more re-enforcing to increase the motivation to find and consume more food. Certainly, pre-civilization human ancestors never contemplated dieting. The concept would have been absurd to them—obesity, too. So, the rub of using starvation for treating obesity is that this solution to a problem of modernity not only lowers the metabolic rate but increases craving and addiction behavior. Unfortunately, this is a relatively nonspecific dopamine response that not only raises the risk for excess eating, but the risk for all other addictions, too. It can worsen a threat-induced imbalance.

Starvation is not sustainable over the long run. Decreased activity, lower metabolic rate, and increased dopamine sensitivity predispose to rebound weight gain and worsening metabolic syndrome once caloric restriction has ceased. Starvation alters the system in expectations of future hard times, increases hunger and foraging, increases the drive for glucose and fructose-filled foods, programs for more glycogen and fat storage, and lowers energy expenditure and muscle mass.

This is ultimately counterproductive. Energy balance, or counting calories-in versus calories-out, is by far too simplistic of a strategy for treating obesity. Through this naivete, real harm can be done. Calorie restriction can induce starvation-threat phenotypes and physiology. Exercise has surprisingly little long-term effects on weight and obesity.[264] School meal programs reduce childhood obesity. Is this because of better nutrition, or perhaps less food and financial insecurity, and less distress within the home? Maybe both.

Many pills and procedures for obesity and metabolic syndrome serve to impair appetite, intake, and absorption. However, these issues are not the source of the problem. To that end, caution should be used when using pharmacologic or procedural disruption or mutilation.

The newer incretin diabetic medications can result in large amounts of weight loss, in addition to controlling blood sugars. The dialogue around the Glucagon-like Peptide-1(GLP-1) and Gastric Inhibitory Polypeptide (GIP) agonists centers on the fact that they reduce appetite. But arguably there may be more going on at a molecular and metabolic, level, and perhaps at a phenotypic level, to account for these dramatic results. There may be a shift in the set point for weight and white fat accumulation to account for their impressive weight loss results. If so, this discounts the theory of excess calories in versus calories out as the primary theory for toxic obesity. Yet, it does not address the primary driver for most toxic obesity—chronic threat.

In fact, GLP1 agonists *do* indeed decrease TS and TC at a transcription level.[265] Perhaps, they work primarily as a biohack to block threat signaling? Interestingly, a side effect appears to be decreased compulsive and addictive behaviors. Trials with GLP1 agonists are being done to look at efficacy in preventing and treating other threat-related disorders, including addictions, infertility, cancers, cardiovascular, diseases, and neurodegenerative diseases including Alzheimer's and Parkinson's disease, and schizophrenia.

Concerns arise around reported side effects that include:

- Nausea
- Vomiting
- Diarrhea
- Constipation
- Anorexia
- Headache
- Nasopharyngitis
- Muscle wasting
- Pancreatitis
- Cancer
- Anxiety
- Depression
- Suicidality
- Emotional and cognitive blunting

These suggest an active TR associated with the medications. Alternatively, could these medications be working through a TR? Could the profound adipose loss and adipokine reduction be reflected in less inflammation, oxidative stress, and metabolic issues, while much of the system is perceiving threat? Or are the medications reducing threat signaling or even increasing safety signaling and an associated SR, with less inflammation and less oxidative stress, along with increased feelings of satiety?

Do these drugs interfere with dopamine signaling to decrease motivation to eat and reward from eating? This would have implications for depressing all dopamine dependent behaviors including addictions—but is this where health and happiness are found? Safety seems like a better solution.

Caution should be used while time tells the rest of this story. The fact that most people regain the lost weight after discontinuing these medications clearly demonstrates that these medications are not curative. They have not gotten to the root—chronic threat.

Chasing elevated blood glucose with insulin in hyperglycemic insulin resistant patients is also folly. Insulin causes more fat storage and worsening obesity and metabolic syndrome. Reducing blood glucose with insulin only treats one symptom of chronic threat, while worsening the underlying physiology. Threat control needs to be prioritized over glycemic control.

In treating obesity, it is important to not create more threat physiology through threat signaling, starvation, blame, shame, guilt, embarrassment, and assumptions of a character flaw. "Gluttony," a sin per Biblical teachings, is seldom the etiology of obesity. Obesity is a phenotype, a consequence of the exposome interfacing with the genome. Lifestyle interventions that focus on the self and self-care strategies may also be detrimental and a set up for failure. Better, interventions should focus on the strategies of treating a chronically high TL for maximum success.

How useful are these classifications in obesity?

- "Hungry Brain" – where the brain does not recognize signals that the stomach is full.
- "Emotional Hunger" – where cravings to eat and what to eat are driven by emotions, anxiety, and negative feelings.
- "Slow Burn" – where people have slowed their metabolism and have a low energy level.

Using the lens of TVST to understand these states as threat phenotypes with programs aimed toward survival may be a better way to understand and manage the complexity of morbid fat accumulation.

The latest data from the National Institutes of Health estimates that more than 40% of U.S. adults are obese, and another 30% are overweight. This means that 7 in 10 adults need to lose weight. Historically, of those who attempt weight loss, only 10% will achieve meaningful weight loss and fewer will sustain the lost weight.

The World Health Organization estimates that, based on a 2017 global burden of disease study, more than 4 million people across the world are dying every year due to being overweight or obese, a trend that is also growing rapidly in children. The recent data suggests that the direct medical costs of a person being obese in the U.S. is more than $2,500 annually. And researchers estimate that the total cost of obesity in the U.S. is more than $1.7 trillion annually, which is 9.3% of the country's gross domestic product. Predictions show the annual cost of obesity will be more than $4.3 trillion worldwide by 2030.

Having a proper etiologic target in treating obesity will help to ease the economic burden of this phenotype. The solutions and the cure to obesity, type 2 diabetes, and metabolic syndrome have less to do with weakness versus strength, or calories-in (eating) versus calories-out (exercising). Rather, they have more to do with threat versus safety.

Additionally, the new curiosity regarding Alzheimer's disease (type 3 diabetes mellitus) possibly being related to the processes of type 2 diabetes mellitus should not be a mystery. Both come from the same root and same soup, chronic threat.

Other Hormones

The focus so far has primarily been on the cytokine peptide signalers, and minimally on GIP and GLP1, smaller endogenous incretin peptide signalers. Yet, there are many other peptide signalers throughout the system. Two others worth mentioning here are *vasopressin* and *oxytocin*, both considered to be hypothalamic hormones. However, they are formed by other cells as well. Vasopressin supports T1 physiology. Vasopressin helps the system hold onto fluid volume, supports blood pressure, is pro-inflammatory, and is associated with aggression. Oxytocin supports S1 physiology. Oxytocin bolsters connection and bonding, along with sexual and reproductive functions. These two small and almost identical peptide signalers have very different functions. One supports threat physiology, and the other supports safety physiology.

Perhaps more importantly, oxytocin appears to be a hormone of self-transcendence. Oxytocin not only supports connection and bonding, but within this transformation there is the loss of the self. This is demonstrated in the awe and profound love a parent has for its newborn and the sacrifices made for the survival of that offspring. Yet, oxytocin can spawn aggression, as seen in the power of a mother's protective response when her offspring is in danger. This response is impressive and can be dangerous for those in her way. Even in this aggressive state the concern for self is absent. Oxytocin has been characterized as the "love hormone," but the "transcendent hormone" may be most appropriate.

Reciprocally, vasopressin is a hormone associated with disconnection, self-protection, and perhaps self-obsession and rumination.

As mentioned previously, in threat, distress and catabolic hormone production are prioritized. In safety, sex and anabolic hormone production are prioritized. States of threat are neither supportive of growth, repair, and regeneration, nor supportive of social, sexual, and reproductive functions.

"True healing is not the suppression of symptoms,
but the addressing of the root cause."

Medical Medium Thyroid Healing

Thyroid function is interesting under the TVST lens, as thyroid hormone production is adjusted to meet the needs of the organism across the spectrum of threat to safety. Thyroid dysfunction and thyroid cancer have been associated with chronic threat states. It is notable that autoimmune thyroiditis has two phenotypes in Grave's disease and Hashimoto's disease. Grave's thyroiditis is a hyperthyroid state, whereas Hashimoto's thyroiditis is a hypothyroid state. These two states correlate well with T1 and T2 phenotypes and thus suggest that auto-inflammatory, auto-metabolic, and auto-immune disease may be phenotypic programs of threat.

Both T1 and T2 thyroid disease is highly correlated with diabetes. Hyperthyroidism is notably correlated with gluconeogenesis and insulin resistance—T1 needs for the fight, flight, or fever. Whereas hypothyroidism is notably correlated with decreased muscle insulin sensitivity and inability of the muscle to use glucose—a T2 strategy to immobilize and store fuel for future hard times. Hyperglycemia is a symptom of a systemic problem brought on by threat and not a disease in and of itself.

"Our fatigue is often not caused by work,
but by worry, frustration and resentment."

Dale Carnegie

Fatigue is a systemic phenomenon. "Adrenal fatigue" has been coined to describe a syndrome of fatigue, body aches, weight loss, hair loss, and lightheadedness as a result of adrenal gland exhaustion from chronic distress. This is an inaccurate description, and there is no evidence of adrenal compromise associated with these symptoms. Just as with "immunoexhaustion" being an inadequate descriptor for the programmed T2 phenotype of immune cells, "adrenoexhaustion" fails to accurately describe what is a systemic programmed T2 phenotype that includes the functions of the adrenal glands, the thyroid gland, as well as the rest of the human organism. The T2 phenotype is a protective programmed response that expressive fatigue but not a state of excess use or using something up, and the T2 phenotype is a coordinated systemic response that cannot be isolated to a specific gland.

Another note regarding hypometabolic phenotypes: the itty-bitty tardigrade, a multi-cellular organism, can alter its morphology and lower its metabolism as a defense mechanism so much that it can even survive in outer space and across extended time. This raises a question as to whether single-celled lifeforms would have this capability—likely—and whether the organic may have been directly implanted, instead of fabricated, on Earth.

"Only humans make waste that nature cannot digest."

Charles Moore

Worth consideration, microplastic pollution is a toxin to the endocrine network much like air pollution is a toxin to the respiratory network. Both need to be addressed. Yet neither microplastic pollution nor air pollution confine their effects to specific tissues as can be seen by their correlations with multiple comorbidities across the entire system. The effects of plastics are systemic from our oceans to our blood to our brains to our glands. It is vital to then move beyond all these siloed concepts to address all threats as systemic and global threats.

Gastroenterology

> *"Trusting your gut. Using your 6th Sense. Having a hunch. These are all ways of describing your built-in intuition or guidance system. We all have it. In fact, your gut makes more neuro-chemicals than your brain. Some of you may feel you have lost your gut-brain, but there are many ways to get it back. Simply desiring to develop your intuition – or regain it – is a good start."*
>
> *Christiane Northrup*

Cytokines and incretins (smaller noncytokine peptide pancreas and brain signalers) regulate appetite, fuel selection, and fuel production, thus determining the status of the gastrointestinal network when in threat versus safety.[266] In threat appetite is suppressed; gastric and intestinal motility slowed; gastric acid and pancreatic enzymes reduced; epithelial tissues inflamed, oxidized, and catabolized; tight cellular junctions are loosened and leaky; and fat is stored in and around the organs of the gut. Threat, TS, TC, and the associated TR can provide a picture of hypometabolism, anorexia, constipation, obstipation, maldigestion, indigestion, ulcers, dysbiosis, and irritable bowel syndrome, if not inflammatory bowel disease, culminating in rising rates of gastrointestinal cancers.[267]

Autonomic parasympathetic regulation through vagal nerve stimulation (notably, the dorsal branch of the vagus nerve primarily innervates the gastrointestinal system and structures below the diaphragm, while the ventral branch of the vagus nerve primarily innervates structures above the diaphragm such as the lungs and heart) shows much promise in chronic gastrointestinal disorders, and other disorders of chronic threat, such as depression, neurodegeneration, respiratory diseases, cardiovascular diseases, fascio-musculo-sketetal diseases (osteoporosis), rheumatologic diseases, some forms of epilepsy, and cancer. [268,269,270,271] The parasympathetic network is the autonomic neurologic cholinergic (where acetylcholine operates as a neurotransmitter) anti-inflammatory and anabolic network for the system that is turned on when in safety. The parasympathetic network directly downregulates TS and TC and upregulates SS and SC.[272]

Arguably, activation of the parasympathetic network through signals or cues of safety by any means—including safe mechanical, electrical, acoustical, ultrasonic, emotional, social, mental, and spiritual therapies—will be helpful tools to use in treating the spectrum of disorders associated with chronic threat. Notably, there is bidirectional flow within the system by which direct stimulation of the parasympathetic network improves emotional, behavioral, social, mental, and spiritual statuses while safe emotional, behavioral, social, mental, and spiritual inputs improve parasympathetic tone and signaling.

Mechanical, acoustical, electrical, or ultrasonic vagus nerve stimulation, or other forms of direct parasympathetic stimulation, appear to be valuable tools to regulate the system. But success, the ultimate biohack, in threat-related disorders will be in removing people from threat and moving them to safety.

The "gut-brain connection" deserves comment, as it has become a very popular topic in both medicine and the general population. TVST contends humans are one system, with multiple networks within the system. The gut and the brain are both part of this system, so yes there is a connection. But it is important to recognize the entire system is connected, too.

In addition, the 40 trillion human cells exist with another 40 trillion colonizing bacteria—the biome. Many of these bacteria are within the gut. All these cells are in communication. They form a bidirectional network influencing each other. Human phenotypes will influence the biome phenotypes, the biome phenotypes influence the human phenotypes, all along the spectrum of threat versus safety. Acute dysbiosis may be primary to pathogens and the threat phenotypes of the biome, but more commonly chronic dysbiosis is secondary to the influences of the threat phenotypes of humans. To demonstrate how connected the system is, contemplate that a cut of the skin will change the phenotype of the gut biome.

Ingested prebiotics and probiotics can influence human phenotypes. However, they do not cure threat-associated secondary dysbiosis, as the human threat phenotype will perpetuate the dysbiosis. Safety will be the most successful treatment for dysbiosis and a healthy gut-brain connection.

Additionally, the concept of "a gut feeling" has some real validity. The older cortical structures of the brain, primarily the cingulate and insular cortices are where the physiology and sensations of the viscera come to awareness. These sensations are not as precise as the sensation from the soma and tend to be a little diffuse, and radiate, or refer, to other areas. It is in this more primitive cortical location where visceral feelings register that emotional feelings also register.

Emotional feelings and visceral feelings run on parallel tracks, with a common middle rail. Visceral feelings come with emotional feelings, and emotional feelings come with visceral feelings. When emotional feelings of threat are repressed from awareness, sometimes the visceral feelings intensify, especially when this is a chronic state. This visceral-emotional phenomenon frequently presents itself in the gut in things such as gastroesophageal reflux disease, gastroduodenal ulcer disease, irritable bowel syndrome, or even inflammatory bowel disease, and potentially cancers. However, the phenomenon presents systemically including within the biome and other human cells, tissues, and organs, as well.

Physical threats are easy for humans to identify. Other threats can be more occult, covert, and insidious for humans to identify. Listening to "gut feelings" helps to discover and sort through and process these other threats. Attending to the signals and communications from the body leads to discovery of and attending to emotions that can help with healing and resilience.

Mental health profoundly impacts inflammatory responses in the body. This is particularly apparent in inflammatory bowel disease (IBD), in which psychological distress is associated with disease exacerbations. Chronic distress, or threat, drives an inflammatory subset of gastrointestinal glia cells within the enteric nervous network that promote TS, TC, and an associated TR. Additionally, TS and TC cause transcriptional immaturity in enteric neurons, acetylcholine deficiency, and dysmotility. There is a direct connection between the psychological state, intestinal inflammation, and dysmotility in IBD patients. The psychologic state, through the brain, directs peripheral inflammation, and defines the enteric nervous network as a relay between psychological distress and gut inflammation. A reduction in the TL is a primary treatment for gut-related disorders, including IBD.[273]

Low gastrointestinal motility and constipation are threat related symptoms that correlate with a risk for dementia.[274] Again, none of these symptoms are causative, but correlative, as they all come from the same root and soup of threat.

An assessment of total TL is essential to understanding human beings. To reiterate again, to feel and be safe, humans need:

- Clean air
- Pure water
- Healthy food
- Access to health care
- Financial stability
- Secure housing
- Expressive freedom
- Supportive social connections
- Healing ideations

- Positive cognitive constructs
- Helpful symbolic narratives
- Reintegration with the natural world

This leads to healing, health, wellness, and thriving.

Truly listening to the gut will help humans to understand and process threats.

Dental

> *"Smiles are contagious – so go forth and contaminate as many people as you can."*
>
> *Ali F. Oeus*

Dental disease—cavities and connective tissue disease (tooth and gum disease and bone loss)—are caused by threat, TS, TC, and an associated TR.[275,276] The biome, or the microbial flora of the body, including the mouth changes in threat versus safety. The phenotypes of human's symbiotic microbial colonizing partners change as a humans' phenotypes change on the spectrum from threat to safety. [277,278] Human cells communicate to microbial cells and vice versa. Dysbiosis is thus more connective and correlative than causative of illness and disease. Chronic distress of the host produces dysbiosis of the gut and other tissues. But at times, the microbiome produces aggressive pathogens and more severe dysbiosis that can inflict an acute illness.

Inflammation, oxidation, catabolism, and cellular changes, including symbiotic colonizing microbial changes, are all hallmarks of a TR. They are hallmarks of dental disease as well. The inflammation, oxidation, catabolism, and resorption of bone, gums, and teeth are hallmarks of a chronic TR. The gums appear to be rich sources of glucose—fuel—for mounting a defense. Periodontal and dental disease are consistent with changes from chronic threat.

Another set of hallmarks of a TR are mitochondrial phenotypic changes and cellular preference for glycolysis. Threat-associated sugar craving and intake further fuels a TR with inflammation, oxidative stress, catabolism, pathologic microbial growth, oral disease, and cavities.[279]

Safety, reduced sugar intake, and improved oral hygiene will prevent dental disease.

Cardiology

> *"If you find it in your heart to care for somebody else, you have succeeded."*
>
> *Maya Angelou*

So too, cardiovascular disease is related to the root and soup of threat.[280] Increased levels of distress and mental health issues are associated with endothelial dysfunction, soft tissue fibrosis and sclerosis, and atherosclerotic disease, and increased neuroticism is associated with arrhythmias.[281, 282,283] The list of comorbidities related to cardiovascular disease is as long as the list of threat-related disorders, including alcohol, marijuana, and opioid use, and addictions. People living in neighborhoods with simply adverse environments that include air pollution, toxins, traffic, lack of parks, and poor access to healthy foods, separate from other determinants such as unemployment, low income, poor education, and crime, suffer from strokes and heart attacks up to twice as often. This all is another reminder that the total TL is at play.

Threat, TS, TC, and an associated TR at an autonomic level will initially deactivate the parasympathetic-cholinergic network and activate the sympathetic-adrenergic network. The hypothalamic-pituitary-adrenal network alters its status with TS, too. These initial threat changes support T1 phenotype needs, including a higher metabolic rate, holding onto fluid volume, elevating blood pressure and heart rate, and inducing more forceful heart contraction.

Eventually, with escalating, compounding, or chronic threat, the T2 phenotype is expressed with deactivation of the sympathetic-adrenergic and hypothalamic-pituitary-adrenal networks, in addition to the previously inactivated parasympathetic network. This results in a lowered metabolic rate, blood pressure, and heart rate, decreased heart contractility, and allows for more autonomous heart rhythms.

This is a picture consistent with fatigue, arrhythmias, and heart failure. Even the effusions and edema of heart failure are explained through TS and TC that initiate loosening and leaking of intercellular tight junctions to allow for defensive protein and cellular migration into the interstitium when in a threat state. But they also result in increased spilling of fluid into the interstitial spaces—swelling, edema, and effusions.

Threat, TS, TC, and an associated TR activate white blood cells, platelets, coagulation factors, and positive acute phase proteins.[284] This not only prepares the system to stop any bleeding from trauma, but these same components of a TR are also participating in the netting, trapping, and killing of pathogens involved in infections.

The inflammatory, loosely packed, low-density lipoproteins transport and allow quick and easy access to cholesterol to be burned as fuel or converted to inflammatory and toxic molecules as needed for a TR. These low-density lipoproteins carry inflammatory phospholipids to needed locations to participate in the inflammatory response to threats. Lipoprotein complexes also contribute to the trapping, and then killing, of pathogens within the blood and interstitial spaces.

This threat-related lipid profile is a hallmark of not only cardiovascular disease, but of all the threat-related diseases, including mental illness. The lipid profile of schizophrenia reflects this commonality.[285]

In a TR, the vascular endothelium acts as a protective barrier for the rest of the system. In threat, the vascular endothelial cells produce high levels of toxic reactive oxygen species and toxic oxidative enzymes to amplify the inflammatory response to protect against invading pathogens. The vascular endothelial cells also alter the lipid make up of their cell membranes and increase their rigidity and stiffness while decreasing the risk for invasion, yet with portals to push out inflammatory chemicals from the cell to guard the cell and create a barrier to pathogens. A fortress with fire outside its walls, so to speak.

However, sometimes in an aggressive TR, there can be collateral damage to cells, tissues, and organs from excess oxidation, inflammation, coagulation, and endogenous toxins.

In addition, fibrocytes change phenotype when in threat and stiffness and thickness increases in connective tissues—this has been coined fibroinflammation. Blood vessels are susceptible to this phenomenon. As a result, this loss of elasticity can lead to high blood pressures, resulting in further endothelial injury, and progressive vascular disease. Arterial stiffness can proceed and correlates with metabolic syndrome, obesity, type 2 diabetes, dyslipidemia…as they all arise from the same root and soup of chronic threat.[286]

The combination of oxidation, inflammation, innate immune cell activation, vascular stiffness, connective tissue thickening, collateral endothelial damage, high blood pressure, clotting, coagulation, and intravascular nets and traps constructed from platelets, chromatin, proteins, lipoproteins, and lipids that were primarily designed to hold down and destroy pathogens, in a response to an aseptic threat—(no pathogen but perhaps emotional, behavioral, social, mental, or spiritual threats and injuries)—can initiate the formation within blood vessels of atheroma, intravascular plaques, and precipitate atherosclerosis. Vascular diseases, such as peripheral vascular disease, coronary artery disease, and cerebrovascular disease, with risks for heart attack and stroke, all follow this threat-related cascade of events. [287,288,289]

Notably, the lipoprotein-cholesterol network is not just a relative risk sideshow to cardiovascular disease, but it is foundational and a major part of the metabolic, immunologic, thrombotic, and even neurologic and psychologic responses along the spectrum from threat to safety. The lipid and lipoprotein network participates in the phenotypes and physiologies of not just defense, but also healing, all along the spectrum from threat to safety.[290] Lipid and endothelial phenotypic changes, separate from vascular disease, correlate with the risk for neurodegenerative dementias—the same root and soup of threat.[291]

Indeed, APOE and beta amyloid have major roles in cell deconstruction and the mobilization of lipids from neocortical tissues to be repurposed in the fight, flight, and fever of a threat response. When glial cells (nonneuronal brain cells) adopt a proinflammatory threat phenotype—coined neuroinflammation, they also participate in neuronal deconstruction and lipid mobilization. Thus, they are seen to have internal fat accumulations. The slippery slope is declaring the lipid accumulation within glial cells as toxic and the source of the disease. This accumulation is simply a phenotypic presentation of threat associated mobilization and repurposing of lipids for the defensive needs at hand.

All threat is catabolic, oxidative, inflammatory, and pro-atherogenic. Chronic threat underpins the spectrum of chronic cardiovascular disease. Chronic threat underpins chronic disease and cancer, and the phenotype of cardiovascular disease is the same phenotype as cancer.[292]

It has been estimated that 50% of all cardiac procedures are unnecessary. Viewing cardiovascular disease through the lens of TVST could further diminish the need for these procedures. In fact, the combination of reduced chronic threat and restored safety and the use of pharmacologic blockers of threat signaling and stimulators of safety signaling could eliminate all cardiovascular disease as it is known today.

All cardiovascular disease is not strictly atherosclerotic disease. There are diseases of catabolism-induced heart failure and diseases of dissolution and autonomy of conduction and pacing in the forms of arrhythmias that can be separate from or coexist with atherosclerotic disease. There are also syndromes associated primarily with cardiovascular symptoms. Again, these all tend to come from the same root and soup of threat.

Postural orthostatic tachycardia syndrome (POTS) is a classic T2c syndrome—faint. Sufferers complain of fatigue, weakness, dizziness, palpitations, and a racing heart when standing. Additionally, syncope, weariness, stomach pain, and headaches are noted. This physiology is demanding POTS patients lie down, hide, be quiet, and perhaps play dead.

Women distressed in their adolescence or early adulthood are the majority of those affected by POTS. Traditional hypotheses are that POTS is either an autoimmune condition, excess sympathetic and catecholamine activity, sympathetic denervation resulting in central hypovolemia, or simply a hyperactive reflex tachycardia. Several autoreactive IgGs have been discovered, including to acetylcholine receptors. Loss of acetylcholine transmission is consistent with impaired parasympathetic function and poor muscle function with loss of strength, and both are consistent with the T2 phenotypes. But for the most part, POTS has remained an unexplained syndrome.[293,294] With all the mystery surrounding POTS, it is noteworthy that increased TL, TS, and TC will induce symptoms entirely consistent with POTS.[295]

Look to the root. TS and TC in the nucleus tractus solitarius of the brainstem are associated with changes in cardiovascular reflexes, hypotension, tachycardia, and multiple organ dysfunction—POTS.[296] POTS is a severe threat-mediated syndrome. POTS is also frequently coexisting with other threat related syndromes—MES, CFS, IBS to name a few. But logically, any severe threat will bring out these commonalities in physiology and these associations. They really aren't different syndromes, as they all arise from the same soup and root of threat, expressed through different cells, tissues, and organs.

A physiologically similar, but acute cardiac syndrome is Broken Heart Syndrome. This is brought on from extreme acute distress and despair, greif, that causes a flood of TS and TC, resulting in a T2 phenotype with an acute (Takotsubo) cardiomyopathy, and in the extreme even death.[297]

Pulmonology

"The breath is the bridge which connects life to consciousness."

Thich Nhat Hanh

The pulmonary network also reflects the phenotypes of threat.

Reactive lung disease, asthma, is a good representation of threat physiology. Threat, especially air pollution, but also food insecurity, poverty, and adverse childhood events including simply living in a neighborhood where gunshots are heard on a regular basis are significant risk factors for asthma.[298,299] Additionally, it is estimated that over 4 million premature deaths occur annually that can be attributed to ambient airborne pollution, most victims live in poorer urban populations. Air pollution is a major risk factor for suicide, as well.[300]

The lungs communicate directly to the brain via the pain pathways to invoke neurologic sickness phenotypes and sickness behaviors when sensing threats. Reduction in neighborhood-related distress and improving opportunity leads to improvements in asthma.[301] Additionally, obesity and TC released from fat cells can directly increase lung inflammation and risk for asthma.

Comorbidities associated with asthma include:

- Adverse effects from medications
- Allergies
- Atopic dermatitis
- Obesity
- Type 2 diabetes mellitus
- Adrenal and thyroid gland diseases
- Auto-immune, auto-inflammatory, and auto-metabolic diseases
- Hypertension
- Cardiovascular disease
- Osteoporosis
- Mental illness
- Cancer

These comorbidities are all also threat-related disorders.[302, 303] Threat compounds threat.

Asthma is a chronic inflammatory airway disease associated with threat, TS, TC, and a TR. TC promote airway mucus overproduction, bronchial hyperresponsiveness, eosinophilia, and immune B-cell IgE synthesis—a hyperactive T1 phenotypic response. However, some asthmatics have few signs of immune activation more consistent with the severe threat associated hypoactive T2 phenotypic response.[304]

Financial, social, and emotional threats have been demonstrated to worsen asthma symptoms in children and adolescents.[305] Lack of safety in the world needs to be a major focus of treating illness, disease, disability, pain, and suffering, particularly in susceptible children.

In fact, socioeconomic deprivation is associated with worse outcomes and increased mortality in all pulmonary conditions.

Air pollution is a toxic exposure with an associated systemic TR that has effects well beyond the lungs. Air pollution has been associated not just with lung disease, but with childhood learning impairments, mental illness, obesity, heart disease, dementia, and cancer to date. Predictably, air pollution, in time, will be associated with all threat-related illnesses and diseases.

In chronic obstructive pulmonary disease (COPD), lung fibroblasts and epithelial cells undergo specific epigenetic changes in a response to chronic toxin exposure. Changes in cellular programs are controlled by a variety of epigenetic modifications. DNA methylation is one mechanism that determines which genes are read and which are not. Transcription factor proteins that activate or inhibit transcription bind to these altered gene locations, affecting the expression of many genes. These threat-related changes are expressed in the phenotype of COPD.[306]

Interstitial lung disease is a chronic inflammatory condition that looks much like a T1 phenotypic response. By contrast, interstitial lung diseases' end stage presentation of idiopathic pulmonary fibrosis appears more like a T2 phenotypic response. In general, fibrosis, whether in the lungs or fascial, synovial, visceral, vascular, reproductive, or breast tissues, is associated with increased and chronic TS and TC, and is more prevalent in T2 phenotypes.[307,308]

Risk factors for interstitial lung disease are exposure to threats, including the toxicity of air pollution, smoke, tobacco, and alcohol.

Pulmonary comorbidities of these lung diseases include:

- Obstructive sleep apnea
- Chronic obstructive pulmonary disease
- Pulmonary embolism
- Pulmonary hypertension
- Lung cancer

Non-pulmonary comorbidities include:

- Obesity
- Diabetes mellitus
- Gastroesophageal reflux disease
- Hypertension
- Dyslipidemia
- Cardiovascular diseases
- Addictions
- Mental illnesses
- Cancers

Again, threat related disorders with elevated TL, TS, and TC.

Pharmacologically altering TC signaling is a major target for disease modification and management in lung diseases. [309] This is certainly reasonable, but safety, SS, and SC may be the most successful treatment.

Dermatology

> *"Skin is the largest organ in the body – semipermeable to the exterosome, it protects us and holds us together. It also lets us know what we are feeling and what we are constraining, repressing, and suppressing through its expression of the interosome."*
>
> *Ali F. Oeus*

Dermatologic disease follows similar patterns to other diseases as outlined within TVST. The skin has the largest surface area interfacing with the environment of any of the tissues and serves as a boundary to the external environment. External irritants, pollutants, and toxins will prompt a TR within the skin. However, many of chronic skin diseases arise from the internal environment and emotional, social, mental, cultural, and spiritual threats, and maladaptive strategies to cope with these threats.

Atopic dermatitis is mediated through threat, TS, TC, and an associated TR. Atopic dermatitis is associated with many comorbidities that arise from the same root and soup of threat. The relationship between atopic dermatitis and other atopic diseases, including food allergies, asthma, and allergic rhinitis, as well as other comorbidities with similar underpinnings, is well documented and understood.

Children of low socioeconomic status are 40% more likely than children of middle and upper socioeconomic status to suffer from allergies and atopic symptoms. Children with atopic dermatitis show lower cognitive performance. Interestingly, in children with allergic presentations, white children are much more likely to outgrow these allergies than black and brown children.

In addition, there is evidence that atopic dermatitis is associated with many non-atopic comorbidities such as cardiac, autoimmune, and neuropsychological (including anxiety, depression, and schizophrenia) disorders, as well as cutaneous and extracutaneous infections, thus establishing atopic dermatitis as a systemic disease, or perhaps better said a chronic threat phenotypic state.[310]

Elevated TC at 2 months of age is associated with atopic dermatitis later in life. Developmental patterns of early atopic dermatitis, allergies, and colic, followed by subsequent diagnoses such as autism, asthma, anxiety, depression, psoriasis, and schizophrenia are found within the past medical histories and problem lists of patients.

Immunologic therapies that directly block TC signaling or inhibit threat enzyme functions and their associated inflammatory and metabolic TRs have recently had great success in treating dermatologic problems, such as eczema and psoriasis. These same therapeutics given for dermatologic issues have been noted to reverse some characteristic symptoms of mental health disorders including schizophrenia and improve all cause survival from a variety of comorbid conditions. All this further supports the idea of a deep systemic physiologic connection to account for chronic disease as postulated by TVST.

Dermatologic conditions also present with phenotypes. The T1 inflammatory phenotype is seen in atopic dermatitis, acne, rosacea, and psoriasis, whereas the T2 fibrotic phenotype is seen in systemic sclerosing conditions. Both phenotypes have an association with not just dermatologic cancers, but other cancers, as well.

TR targeted immunologic and metabolic therapeutics can block TS and TC to reduce symptoms but cannot fix chronic threat nor cure threat related disease. Pharmacologic biohacking of the healing SR may have promise for future treatment, but success will be found in reducing threat and increasing safety physiology through more holistic and comprehensive treatments. Understanding total TLs will allow the design of treatment algorithms to cure—at least until threat reigns again.

Rheumatology

> *"Some days there won't be a song*
> *in your heart. Sing anyway."*
>
> *Emory Austin*

Rheumatology is the study of inflammatory and immune diseases. Auto-immune diseases are highly correlated with and mediated through threat, TS, TC, and the associated TR.[311] Immune-related diseases correlate with dementias and other threat related disorders, as they arise from the same root and same soup of threat.[312] A simple diagnosis, such as chronic sinusitis, can be a signal of a high TL and even a prodromal symptom to rheumatoid arthritis and other threat related disorders.

Given the pleiotropic nature of cytokines, auto-immune disorders need to be considered not only auto-immune, but also auto-inflammatory, and auto-metabolic disorders. Indeed, there is a high correlation with TL and these "auto" diseases. Adverse childhood events and excessive adult distress are well correlated with autoimmune diseases.[313]

Multiple sclerosis, a neurodegenerative demyelinating disease, has recently been correlated with adverse childhood and adult events, and elevated TLs, suggesting that it too is a threat-related disease.[314,315]

Other common comorbidities seen with MS include:

- Anxiety
- Depression
- Hypertension
- Dyslipidemia
- Chronic lung disease
- Cardiovascular disease
- Coagulation disorders
- Auto-immuno-inflammo-metabolic diseases (i.e. psoriasis, rheumatoid arthritis, inflammatory bowel disease)

The question of whether multiple sclerosis is an auto-immuno-inflammo-metabolic or a neurodegenerative disease is answered with a "yes"—all the above. Fundamentally, MS is a threat-associated disease.

In MS, phenotypic structural and metabolic changes can be observed prior to inflammatory changes.[316] Too often chronic diseases are only described in terms of inflammation, and immunity to a lesser degree, to the omission of the degenerative and catabolic changes associated with these phenotypes. All phenotypes have parallel immune and metabolic systemic responses. Threat phenotypes are characterized by inflammation and catabolism with either hyperimmune (T1) or hypoimmune (T2) responses.

Demyelination, or loss of the insulation around neurons, is not uncommon in other threat states. Such states include migraines and severe infections, including Covid-19.[317,318] Migraines can be a prodrome to MS, and seizures can be a late effect within this phenotype. Myelin can be a rich source of fuel for the fight and flight. It is high in sugar and fat, with galactocerebroside being myelin's major structural component.

In models of MS, microglia cells have specific phenotypic changes consistent with a TR that participate in the repurposing, deconstruction, and destruction of neurons. The expressed transcriptomic signature is indicative of TC regulation, with associated reactive oxygen species production, altered lysosomal functions, and metabolic changes all paralleling ongoing phagocytosis. Microglia serve multiple roles during demyelination and their transcriptomic state resembles other neurodegenerative conditions. Phagocytosis is likely a universal phenotypic activity common to all neurodegenerative microglial threat states.[319]

In addition, a recent study demonstrated that holding immune cells in a safety phenotype and promoting SC production can heal demyelination.[320] Infused recombinant safety cytokine interleukin-4 (IL-4) and granulocyte colony-stimulating factor (G-CSF), upregulate healing mechanisms and produce an array of growth factors, thereby gaining the capacity to promote neural growth and regeneration. Moreover, transfer of

IL-4/G-CSF-polarized bone marrow neutrophils into experimental models of neural injury trigger substantial axon regeneration within the optic nerve and spinal cord. These findings have far-reaching implications for the future development of therapies that may bring us closer to effective solutions for reversing neural damage.

Additionally, safety cytokines are capable of signaling regulatory lymphocytes to suppress neurotoxic astrogliosis in the brain, drive alveolar regeneration in the lung, increase angiogenesis in the limbs, decrease renal and hepatic inflammation, and enhance corneal healing in the eye. Immune cells transition phenotypes from defensive, degenerative, and destructive properties to healing, regenerative, and reconstructive phenotypes in threat versus safety.

MS treatments have focused on blocking the immunologic defense response by sequestering or inhibiting immune cell function. This can slow progression of the disease, but will not heal, as the safety phenotypes of these immune cells are required in a healing response. TVST predicts this. To that end, it suggests that a focus on pharmacology that also enhances safety physiology to complete a healing cycle is needed in research and development across all fields of medicine.

More specifically, myelin is produced by oligodendrocytes in the central nervous system. Myelin is the insulating sheath that surrounds neuronal axons that allows for faster conduction velocities along nerves, a huge evolutionary advantage to be able to communicate quickly across large organisms. This evolutionary innovation, which first appeared in jawed vertebrates approximately 500 million years ago not only enabled rapid transmission of nerve impulses, but allowed for more complex brains, greater morphological diversity, and increased size of organisms.

Interestingly, this ability to manufacture myelin seems to be tightly tethered to a retroviral origin being expressed within human DNA. This retroviral contribution to the human genome seems essential for myelination formation, suggesting that retroviral endosymbiosis played a prominent role in the emergence of vertebrate myelin and the human species. Demyelination appears as a combination of a metabolic and a hyperactive immune/inflammatory response. How much of this is driven by this retroviral connection and perhaps an associated increased antigenicity of myelin is TBD.[321]

"I find it astonishing that the immune system embodies a degree of complexity which suggests some more or less superficial though striking analogies with human language, and that this cognitive system has evolved functions without the assistance of the brain."

Niels Kaj Jerne

TVST suggests that antibodies are involved in the coordination of the entire system. In the case of antinuclear antibodies, the immune network is integrating TS through selective nuclear threat expression. The result is catabolism and inflammation of various tissues and organs.

Antinuclear antibodies do not seem to be specific nor diagnostic to auto-immune diseases and are somewhat ubiquitous. However, they do seem to be correlative to threat and an elevated TL.[322] Case in point: elevated antinuclear antibody levels in nonauto-immune illnesses such as Covid-19.[323] A worrisome statistic is that antinuclear antibodies in humans have increased by 50% in the last three decades. The current attribution is to bad health habits, pollution, processed foods, infections, climate change, and stress. TVST suggests an even more globally high TL to be the cause.

The auto-antibodies of myasthenia gravis (a condition which results in the abnormal weakness and fatigue of specific muscles) interfere with the signaling of movement in the body and perhaps reflect the antibody network signaling of an immobilization T2 response. In the genetically predisposed, could this disease be programmed to appear in some individuals in chronic threat states? There is a strong correlation between myasthenia gravis and depression.[324] They likely come from the same root and soup, chronic threat.

Additionally, antibodies play a role in healing processes.

Auto-antibodies are notably involved in the healing response from a Covid-19 infection. Those with specific types of antibodies recover quicker and are less likely to suffer from a post Covid-19 syndrome, otherwise known as Long Covid. [325] Indeed, antibodies may have both threat and safety phenotypes, paralleling TC and SC, and be more than just mediators of an immunity and a TR.

As with cytokines, TVST suspects pleiotropic properties to antibodies that extend well beyond their canonical immunologic properties. This may include transcription, translation, and metabolic properties.

"On Suffering with Gout
Alcohol is the worst enemy
Water is the best friend"

William Tibbles

Gout has had associations with fructose, alcohol, rich food, mammalian meat, high purine content organ consumption, and interestingly, historically with excess wealth. Gout is considered more of an inflammatory disease than an immune disease, as auto-antibodies are not present. But with gout both the immune and metabolic networks are very much involved.

Gout is caused by urate crystallization within joints which occurs when serum uric acid levels are too high. With gout, the precipitation of uric acid as needle-like urate crystals in the soft tissues of the joints is just part of the problem. Inflammation and oxidation are systemically high and most acutely represented within the joints. This combination of crystal deposits, inflammation, and oxidation cause an erosive inflammatory arthritis, which is initiated and sustained by mechanisms involving multiple pro-inflammatory mediators, such as TC—peptide signalers. However, purine threat signaling is the more notable in gout. The purine, extracellular adenosine triphosphate (eATP), is degraded into other purine metabolites such as xanthine, hypoxanthine, and uric acid. An increase in the breakdown of the purines has correlates with elevations in ammonia, uric acid, inflammation, and oxidative stress, and gout.

Also associated with this pathway are metabolic syndrome, cardiovascular disease, liver disease, kidney disease, preeclampsia/eclampsia—i.e., this pathway presents itself in threat related illness and diseases. This makes sense, given that eATP may be the most primitive and primary signaler of threat. eATP levels will be high in threat physiology and eATP must be metabolized, therefore uric acid levels will rise in threat.

Gout is not just the precipitation of uric acid crystals and the associated arthropathies (joint diseases) from too high a dietary intake of purine-laden foods. It is also a symptom of systemic threat, TS, TC, and an associated TR. Uric acid is a metabolite of threat physiology, and gout responds to many different medications that intervene in the associated immune-inflammo-metabolic cascade of threat. This includes TC inhibitors, purine targeted drugs, and steroidal and nonsteroidal anti-inflammatories drugs.

The historical connection of gout to the wealthy may have to do with the consumption of delicacy foods such as pate foie gras, liver, or other highly prized organs. These organs are high in purines, raising the risk for gout attacks—not to exclude an aristocratic sedentary lifestyle and alcohol consumption. To produce really good pate foie gras, animals are immobilized and force fed. They are pushed into obesity and metabolic syndrome, and most of all prized for their fatty liver disease—the main ingredient in pate. To some degree most corporate and cage raised animals suffer from this syndrome. High purine and high fat (adipocytes filled with TC) consumption is a setup for gout.

> *"Gout: A physician's name for*
> *rheumatism of a rich patient."*
>
> *Ambrose Bierce 1842*

With regards to mammalian meat intake, it is worth considering whether ingesting meat from closely related relatives such as cows and pigs that are filled with terror cytokines at the time of slaughter could result in the adverse health implications (including precipitating gout, but also cardiovascular disease and cancer) correlated with mammalian "red" meat intake. TC are active in the intestines, but typically from internal production. How well these peptides survive the acids and enzymes of the upper oroesophogastrointestinal tract is TBD. However, the possibility that ingesting TC is a source of disease, disability, pain, and suffering deserves further consideration.

Hematology

> *"The nitrogen in our DNA, the calcium in*
> *our teeth, the iron in our blood, the carbon*
> *in our apple pies were made in the interior*
> *of collapsing stars. We are made of starstuff."*
>
> *Carl Sagan*

The study of the blood includes the examination of white blood cells (commonly referred to as leukocytes or immune cells), red blood cells, platelets, plasma, and the various peptides and proteins carried throughout the body by the blood. The blood cells are formed within the bone marrow. Blood contents are primarily carried within the vascular network, including the heart, blood vessels, and lymphatic tissues and organs.

Evolution of the vascular and neurologic networks occurred with the evolution of larger multi-cellular organisms around 500 million years ago. The neurologic network can communicate and regulate the greater system rapidly across distances via electrical conduction and neurotransmitter signaling. The vascular network, via the circulating of the blood (peptides, proteins, hormones, cells), can also communicate and regulate the greater system across distances, although at a slower pace. It is notable that the mechanisms for immunoneurology evolved within the nearly 4 billion years prior to the evolution of multi-cellular and differentiated-cellular life and vascular and neurologic networks. These mechanisms have been conserved in humans.[326]

Blood is technically a form of a connective tissue, with a mobile liquid structure containing highly mobile cells that reach to all the corners of the body.

The white blood cells are typically seen as the system's defense network. However, as seen through TVST, the entire system assumes a role in the defense of the system. The "immune" cells are the most deployable and mobile cells within the system, and they can quickly get to a site of threat. These cells are unattached and have more independence than most cells in the human body. In their morphology and function, they resemble ancient eukaryotic single-cell organisms more than any other cells in the body. These cells in a defensive mode are also the best armed and equipped to destroy a threat.

Like all cells, the blood cells take their lead from the environment. When in threat they are aggressive, inflammatory, and destructive in nature. When safety is restored, these same cells reduce inflammation and participate in resource distribution, reconstruction, and healing efforts. This includes cleaning up debris, recycling cellular materials, and deactivating, deconstructing, and destroying dysfunctional cells.

Immune cells also have a phenotype associated with advanced age and chronic threat consistent with a T2 phenotype. This has been previously characterized as "immune-senescence," "immune-paralysis," or "immune-exhaustion," in which the immune cells are depressed and do not function, defend, or heal well. The immune cell phenotype changes with the ever-changing environment from threat to safety.

People with hedonistic traits—pleasure seekers, users, and abusers—interestingly have relatively elevated inflammatory markers and suppressed viral immunity. In contrast, people with eudaimonic traits—producers, improvers, and connectors—have relatively lower inflammatory markers and higher viral immunity. This seems to represent a wonderfully intelligent phenotypic shift in immune cells. Those with threat-associated traits tend to become more isolated and are at risk for injury from bacteria and predators, thus requiring a more aggressive inflammatory defense. But those with safety-associated traits gravitate to sociality and proximity to others. Viruses are dependent on the congregation of people to spread, and the immune system predicts this threat and supports viral immunity when humans are more social and safe.[327]

Consider that when isolated within the Covid-19 pandemic, viral transmission was controlled, the spread of Covid-19 was slowed, but both individual and population viral immune functions were lowered, while individual and population TL, TS, and TC levels were elevated. With lowered viral immunity and higher inflammatory physiology this became fertile ground for the virus and more severe illness with worse inflammation including cytokine storms, worse outcomes logically followed.

Those who became infected would have a more difficult course due to lower viral immunity and from having moved their physiology closer to the critical threshold of a cytokine storm that causes massive cell, tissue, and organ injury, as well as death. Covid-19 morbidity and mortality are associated more with the level of TL, TS, TC, and the associated intensity of a toxic TR than with the viral load itself. It has been demonstrated that the viral infection and associated brain injury is not related to the viral load, but to the intensity of the immune, inflammatory, and metabolic response to the pathogen—all set by the total TL, TS, TC, and the associated TR.[328]

In the case of cytokine storms, the white blood cells produce massive amounts of TC, resulting in T2c phenotypes, fainting, shock syndromes, cardiovascular collapse, and even death. Many of the TC are cell interleukins, as these peptides were first discovered in association with white blood cells, leukocytes. It is now known that cytokines are produced by almost all cells in the body, not just immune cells, and that they do far more than produce inflammation and an aggressive immune response. TC activate a systemic threat response that includes metabolic, or catabolic changes, for fuel to stoke the threat response. This, in combination with the collateral damage from the inflammation and immune attack, accounts for the severe cell, tissue, and organ injuries associated with shock and Covid-19.

Also consider this: hedonia is a self-focused state. Anytime humans are self-focused, threat physiology and a defensive state may be assumed. Eudaimonia is an other-focused state. Safety allows us to move beyond the concepts of self. In addition, consider that safety phenotypes have high surveillance and viral immunity, and surveillance and viral immunity has a correlation with cancer immunity.

White blood cells are predictive in nature, and they have memory capabilities to go with this function. White blood cells are called immune cells for their ability to defend and remember. It is the ability to remember that gives learned immunity. Exposure to previous threats, and particularly previous antigens and allergens, create a memory or a code to be able to quickly recognize and neutralize a threat in the future. B lymphocytes produce antibodies to amplify this learned response.

White blood cells and cytokines are indeed heavily involved in both immune and metabolic regulation, as a defense cannot be deployed without a fuel strategy to support it. White blood cells in a defensive mode convert to glucose as a fuel source and glycolysis for energy production. Interestingly, metabolites of glycolysis block tumor suppression and facilitate the initiation, propagation, and metastasis of cancers. Glycolysis is a primitive but fast, although less efficient, metabolic mechanism to produce energy. This connection and interdependency of immunity and metabolism is a significant tenet of TVST. (Another tenet is that defensive functions go way beyond the cells of the blood, but the cells of the blood do have major roles.)

Any model of threat-associated disease must incorporate both immunity and metabolism, as they go hand in hand, within a systemic response to threat.

White blood cells are mobile and can escape the confines of the blood vessels to attack invaders in the interstitium. They can net and trap foreign materials and invaders. White blood cells will sacrifice themselves at times for the greater good. When netting a threat, white blood cells can spit out their DNA/chromatin and use histone proteins within the construct of a net to not only amplify a TR, but to trap a threat. In this regard, a white blood cell is a form of an altruistic suicide bomber assigned to protect and propagate the genetic code.[329]

Some white blood cells have phagocytic properties that can be deployed to destroy a pathogen. They also have toxic programs by which they can attack a threat with enzymes, oxidants, and acids. They are potent signalers of threat and can readily amplify TS and a TR throughout the system. Cytokine signaling is one of these mechanisms.

The white blood cells, like their cousins, the glial cells of the central and enteric neural networks, have some regulator and structural controls as well. These cells decide if other cells and tissues will be constructed, deconstructed, or destroyed. The white blood cells can survey for debris, pathogens, and abnormal cells, and remove them by various means.

One question with cancer cells is are they abnormal cells? Or are they simply a phenotype of the self? It is likely they can be both. Immune cells are more likely to remove abnormal and weakened cells but leave the phenotypically altered cells of the self alone. The cancer cell may not be evading the immune response, as much as just being the agitated and threatening form of the self that is tolerated by the system, at the system's own peril.

"Inside each of our 25 trillion red blood cells are 270 million hemoglobin, each of which has room for four oxygen molecules. That's a billion molecules of oxygen boarding and disembarking within each red blood cell cruise ship."

James Nestor

The red blood cells transport oxygen to the other cells in the body. The oxygen can be used by the mitochondria for respiration and oxidative metabolism to produce voluminous amounts of energy to run the functions of the cell. Or the oxygen can be used to produce toxic reactive oxygen species or oxidative enzymes as part of a defense. The red blood cells have no nucleus or other internal organelles. These structures were spit out of the cell in its youth. The red cells are dependent on glycolysis for energy production. They cannot reproduce on their own. The lack of internal structures allows them to carry more oxygen and be smaller and more flexible to get into every nook and cranny of the body. It also means they have no value to viruses in terms of viral reproduction. Red blood cells will produce eATP (purine signaling) when they sense threat, but they have no mechanism by which to produce peptides or proteins within a threat response. If destroyed, the free hemoglobin and abundant oxygen released amplify inflammation and a TR.

Platelets are really a particle of a precursor cell. Platelets do not have internal organelles. This makes them smaller and more flexible, thus improving their functionality. They also have phenotypes—a spheroid and calm one in safety and an arborized and mobile one in threat. Platelets in a threat phenotype are looking to attach and connect things together. They can thus stop bleeding. They also net and trap debris and pathogens. This immobilizes pathogens and allows the white blood cells to rally and destroy the invaders. Platelets also sense threat and signal threat, but they don't have the infrastructure to produce peptide or protein signalers.

The blood also contains noncytokine larger signalers and coordinators of a TR. Acute phase proteins include:

- C-reactive protein
- Amyloid
- Complement factors
- Fibrinogen
- Prothrombin
- Factor VIII
- Von Willebrand factor
- Ferritin

They are all coordinated to amplify the TR, stop bleeding, trap pathogens, and ultimately, to start healing.

The human defense system was designed to protect from and fight off pathogens and predators. When the threat is not a predator or a pathogen, but chronic emotional, behavioral, social, financial, mental, or spiritual threat, the response can be very similar. The reason is these blood cells assume a threat phenotype with these threats, too. The outcome is persistent inflammation and catabolism, with tissue pain, fibrosis, and degeneration, if not frank auto-immuno-inflammo-metabolic diseases. In addition, the processes of netting and trapping debris and pathogens are very similar to the process of atherogenesis—the building of inflamed white cell, platelet, fibrin, lipid, chromatin, and calcium plaques within the blood vessels. Useful acutely, harmful chronically.

These are maladaptations of chronic threat, but the system is doing exactly what it is supposed to do when sensing threat. The chronicity of threat physiology is the problem and the detriment to the organism over time.

An acute TR is a protector. It is a friend, not an enemy. Also, almost every aspect of a TR response can be modulated pharmacologically, but every modulation disrupts normal physiologic processes. Safety shuts down the TR to prevent the maladaptations of chronicity and promote healing without disrupting or destroying the normal functions of the TR that humans need to survive when future threats arise. The components of the TR are not the correct targets for wellness and health, the chronic threats are.

Oncology

"Yesterday is history, tomorrow is a mystery, but today is a gift—that's why it's called 'the present.'"

Eleanor Roosevelt

The World Health Organization (WHO) has recently highlighted the growing burden of cancer across the globe and the striking inequities effecting the poor. The incidence of cancer went up 50%, and cancer deaths increased 18% between 2012-2022. By 2050 there is predicted to be another 77% increase in the cases of cancer, with new diagnoses reaching over 35 million. This will disproportionately affect the poor at nearly double the rate of increase being seen in this population. Approximately 20% of all people will develop cancer in a lifetime, and 10% will die from cancer. Cancer care is expensive, topping over $200 billion dollars annually.

"As long as, there are flowers and children and birds in the world, have no fears, everything will be fine."

Nikos Kazantzakis

Perhaps even more concerning are the rising rates of cancer in our young. For example, colorectal cancer rates in Americans have risen 500% in ages 10-14, 333% in ages 15-19, and 185% in ages 20-24 from 1999-2020.

Elevated threat, TS, TC, and an associated TR lead to the induction of cancers.[330,331] Therefore, great consideration needs to be given to cancer within this model of threat versus safety.[332,333] The model historically used for cancer is similar to that used for allergens, antigens, or pathogens. In the case of cancer, it is a search for a mutation, an abnormal gene, even an abnormal nucleotide within a gene, as the source of an abnormal cell and as a potential target for a cure of cancer. It is as though cancer comes from an isolated exposure or a single, isolated, solitary abnormality. It is as if cancer is a "cancogen." Yet, it is clear that cancer phenotypes can be expressed via epigenetic changes without genetic changes. Arguably, these epigenetic changes should be reversible, and this has now been demonstrated. Epigenetic drugs are currently being developed for the treatment of cancers. The traditional model for cancer has been at least partially flawed.

With cancer, a current focus is to use the immunologic network to identify cancer cells, "cancogens," and eliminate them, much like the model for removing pathogens. Immunologics, such as targeted monoclonal antibodies, juice up this response. While metabolics, such as kinase inhibitors, can interfere with the basic metabolic and reproductive functions of the threat-phenotype cells called cancer, and reduce threat cytokine production. But noncancerous cells depend on these pathways, too, leading to unwanted and sometimes toxic side effects including infections and other cancers. Chemotoxins are also used in the hope to preferentially destroy rapidly dividing and replicating cancer cells over noncancer cells to afford a cure. At times, this looks like a race to the death of the two lineages of cells. The threat of chemotoxins appears to produce another or compound the threat code and place patients at risk for future cancers.

Admittedly, there are some successes within this model of care. Although cancer rates are escalating, more cancers are being cured than ever before. Yet, chronic threat, TS, TC, and the associated TR will continue to cause genetic and epigenetic changes to DNA, histones, and RNA that result in the increased expression of onco-genes (pro-cancer genes) and onco-phenotypes (the phenotype, as opposed to the pathotype, of a cancer cell).[334] In cancers, general global DNA hypomethylation is accompanied by selective focal hypermethylation in certain regions of the genome. While global hypomethylation can induce the expression of onco-genes, focal hypermethylation can suppress the expression of tumor suppressor genes. Collectively, these processes can lead to the initiation, propagation, and metastasis of cancer.[335]

Chronic threat influences both the genetic and epigenetic

changes in the phenotype of cells that can initiate, differentiate, propagate, and disseminate cancer cells.[336] The mechanisms can already be seen by which threat influences transcription of the genome to change the phenotypes of cancer cells to become more aggressive, invasive, and metastatic.[337]

Recently, it has become clearer that toxins, pollutants, and perhaps viruses, do not necessarily cause direct mutations of DNA. Rather, they stimulate threat related genetic and epigenetic changes and induce onco-genetic expression within certain cells to present as independent and autonomous cancer phenotypes. Both high and chronic cortisol and TC levels can initiate and propagate cancers and promote metastases. This is consistent with the expectations under a threat associated dissolution process within TVST.[338] Cancer is an extreme threat-associated dissolution all the way to the level of the cell with cellular infidelity to the whole.

In addition, cancers change and evolve over time. They are not fixed and immutable. They fortress within a tumor (T1a), can become more aggressive (T1b), and even run, or metastasize (T1c), thus spreading cancer cells around the body. As they adapt and evolve, they tend to get better at evading the immune system. Immature cancer cells and less aggressive cancer cells can use oxidative phosphorylation. But as the TL increases and the cancer advances, a more glycolytic phenotype is the norm. Associated with these changes, cancer cells exhibit more aggression, autonomy, infidelity, and mobility. [339] A tumor starts as a single, corrupted cell, but becomes a mixture of millions of cells that have changed slightly but in many different ways.

Tumors are adaptable and have almost infinite possibilities to which they can evolve. Cancer is a disease which evolves as it progresses. Late-stage cancers can become very hard to treat successfully. Given these realities, prevention becomes preferable to chasing the disease.

Distressed mitochondria will push their DNA/chromatin through their membrane into the cytosol of the cell. This stimulates an inflammatory response—a TR. (Extrachromosomal and extracellular DNA are known signalers of danger (DAMP) and increase a TR.) In addition, this free extra-mitochondrial DNA can enter the nucleus of the cell to be incorporated into the genome of the cell to further threat phenotype expression and threat physiology, and arguably promote cancer phenotypes. One also must consider whether this process of mobilizing DNA is a last-ditch attempt to try and save the genetic code within a suffering organelle, to more rapidly transcribe and translate the code, to signal danger, or to directly alter the code—the genotype and phenotype—of the cells.[340] Arguably the answer is—"yes."

In addition, this free mitochondrial DNA can be transferred into the nuclear genomes of eukaryotic cells to become embedded much like a viral insertion. This transfer has been linked to shorter lifespans. Within the human brain, insertions of mitochondrial DNA into the nuclear genome are much more common in the "sapiocortex" compared to other brain regions, and the number of insertions may accumulate over time. This all suggests these are not random insertions, but a programmed response to distress. The implications from these mechanics may be huge in the area of not just cancer research, but dementias, as well.

Distressed cells also demonstrate the spitting out of their DNA/chromatin from the nucleus into the cytosol. Extrachromosomal DNA can aid cancer growth by harboring onco-genes that boost the efficiency of transcription and the inheritance pattern of this DNA, resulting in many copies in a single cell. This supports rapid amplification of the onco-gene burden, tumor growth, and tumor evolution.

Although present in many types of human cancer, extrachromosomal DNA was thought to be non-existent in normal tissues. However, extrachromosomal DNA was recently found in phenotypically changed non-cancerous tissues known to be predisposed to cancer phenotype development. Extrachromosomal DNA may simply be a sign of a distressed cell and phenotypic change in non-cancerous tissue, to later become cancerous. It may signify a correlate, but not the cause, of the cancer.[341] Extrachromosomal DNA thus should appear before the infidelity and autonomy of cancer forms.

This similarity to the functions of distressed mitochondria raises the question as to whether the nucleus, like the mitochondria, is also a remnant of an ancient endosymbiotic microbe. The membrane around the nucleus may not be an evolutionary adaptation to protect the code from cellular digestion, but simply the membrane of an endosymbiotic microbe that was able to prevent destruction by its host cell to coexist. Nevertheless, is the spitting out of the code a conserved program to signal danger? Or is it an attempt to save the code when in distress? Or is it simply a mechanism for more rapid transcription and translation of the code? It appears at least to be a marker for threat.

This gives consideration to cell-free DNA (cfDNA) and circulating tumor DNA (ctDNA)—measured in the serum for liquid cancer biopsies—as being more than just free DNA because of cell destruction. Rather, the presence of ctDNA may be a result of a programmed process of spitting out the DNA for signaling danger to other cells or to save the genetic code prior to succumbing to the perceived toxic and destructive environment. That threatening environment can include the metabolic, immunologic, or chemotherapeutic treatment interventions.[342,343] If threat induced cfDNA or ctDNA can be taken into the cytosol, then into the nucleus, then spliced into the genome of other cells this has tremendous implications not only for cancer and the mechanisms of "metastases," but for all illness, disease, disability, pain, and suffering.

Metabolically, cancer cells tend to express primitive threat phenotypes with a reduction in pH, reduction in anti-oxidant and electron transport chain enzymes, impairment in oxidative phosphorylation, increase in oxidative stress with toxic reactive oxygen species and oxidative enzyme production, and a glycolytic metabolic preference.[344,345]

Behaviorally, cancer cells employ common defense responses as if under threat. They colonize (tumor) and fortress, armor, protect, and defend within the colony.[346,347] Tumors are indeed colonies, not just clumps of cancer cells. Tumor cells strategize, specialize, and socialize to protect and propagate the new cell line—the new code.[348] These cells also fight, become aggressive, attack, and invade other cells, tissues, and organs. At times, they appear to run in the form of metastases.[349] Cancer cells can become asocial and be resistant to apoptosis, the altruistic form of programmed cell death.[350] Perhaps they spit out their genetic code in a survival strategy, as bacteria do, too. And, as their primitive bacterial and archaeal ancestors did, they even reproduce rapidly as a survival strategy.

Cancer cells show all these phenotypes of threat. There will be genetic clusters and patterns, hidden patterns to us right now, that will be discovered and defined in time. The lens of mechanistic adaptation may be more useful than the lens of random mutation to see the processes of cancers. Even with our evolving knowledge of cancer ultimately, it may be the understanding of TVST and the proper treatment of threats with the restoration of safety that may finally triumph in the cure of cancer.

"Cancer must feel like such a betrayal, knowing that somewhere deep in your body you're manufacturing tiny bombs that detonate and catch fire."

Mary H.K. Choi

TVST postulates most cancer is not a disease of the cell; it is a phenotype of the cell, yet a betrayal to the whole. Therefore, look to the environment of threat to find the root cause of this infidelity to the whole and a cure. It is worth acknowledging that many humans today exist in chronic threat, a chronic threat phenotype, and infidelity to the whole—a cancer phenotype. Threat multiplies, accumulates, compounds, grows, and metastasizes if left unchecked from a cellular to a global level.

Many major questions that remain unanswered:

- Are cancer cells simply fighting to survive in what is perceived as a toxic environment?
- Are cancer cells advanced threat phenotypes arising from the genome?
- Is perceived damage, error, and mutation really a random event in need of repair or elimination, or some other process?
- Are cancer cells truly mutants and genetically "abnormal," or are they threat-programmed genotypic and phenotypic cell variants?
- Are cancer cells the expression of chronic threat, excessive threat load, and the dissolution of control to the cellular level?
- Is cancer a strategic modification and expression of the genetic code, resulting in a cellular threat-survival response?
- Have cancer cells strategically lost the social cooperation and connection to, and the control from, the greater system?
- Are cancer cells the ultimate expression of a cellular threat phenotype—autonomous, aggressive, inflammatory, degenerative, rigid, inflexible, and antisocial—that stand in contrast to stem cells (pluripotential progenitor cells from which all cells evolve and differentiate)

as an expression of a cellular safety phenotype—civil, cooperative, regenerative, adaptive, flexible, and social?

- Do stem cell populations fall and cancer cell populations grow in threat phenotypes?
- Why are genetic changes across cancers and tumors so similar if they aren't preprogrammed genetic code activated into a cancer cell phenotype?
- Can the genome, under the influence of the exposome interfaces, change itself in threat versus safety? Does the code write its own new code? How plastic is the genome?
- Can threat signaling dictate strategic nucleic base, nucleotide, changes to the genome?
- Does threat induce "jumping genes," transposable elements (TEs), to change the genome itself into a threat or cancer phenotype?[351]
- Are these mutations threat-related programmed functions of the genome to change its own genotype?[352]
- Would the chronic threat-induced turning off the intercellular safety-collaborative genes and the turning on the suppressed primitive threat-autonomy genes be where cancer transformation can be found?
- Would reversing this signaling be where the cure for cancer lies?
- Are the genes for cellular cancer transformation found buried within the evolution history of genes that allowed for cell specialization, cell differentiation, and complex coordinated multi-cellular life approximately 500,000 million years ago?
- Do cancer treatments need reconsideration?
- In cancer care, can the antitumor aspects of the acute inflammatory and catabolic innate immune response be used on a more short-term basis?
- Then can a rapid transition to a short course of an activated adaptive immune response be invoked?

- Followed by the stimulation, initiation, and sustenance of the anti-inflammatory, anabolic, surveillance, and healing immune response to complete a healing cycle and restore the cells to a safety phenotype?
- Does cancer treatment require the entire spectrum and continuum of TS to SS to not only defend the system, but to heal the system?
- Do therapeutics that signal more threat used long term create an environment conducive to and compounding of inflammation, oxidation, catabolism, mutation, onco-gene expression, onco-phenotypes, cancer cells, tumor initiation and growth, along with cancer invasion and metastases?
- Is the cure for cancer to be found in dissipating threat, TS, and TC i.e. the TR and inducing safety, SS, and SC i.e. the SR?
- If given the proper signals, do cancer cells have the genetic flexibility to transform back to more civil, connected, social, and less aggressive, less autonomous, and less antisocial phenotypes (stripping epigenetic coding from rhabdomyosarcoma cancer cells has shown this ability to undergo a phenotypic change back to stem cells[353])?
- Is it possible to ask an organism to be a good citizen while waging war on that citizen to kill it?
- Does the cancer bring the war, or does the war bring the cancer?
- Is fighting a war on cancer a failed paradigm?
- What if cancer cells need to feel safe to change phenotype and to "cure" cancer?
- Finally, what if cancer is not so malevolent? What if, as the vines in the forest pull the diseased and the fallen back into the earth to be recycled to life again, cancer cells do the same for the human body?

These are all reasonable questions today that may have been laughable yesterday.

All the answers to these questions have yet to be determined. For the moment, prevention through preemptively reducing threat and increasing safety should be a focus in cancer care, all of health care, and for humankind.

Cancer is a manifestation of threat, and threat behaves like cancer in that it multiplies, accumulates, compounds, and metastasizes to become an infidel, not just in the physical body, but within the body politic, institutions, society, and culture.

Gerontology

> *"Death feeds life. The snags, the stumps, and the fallen create the lushness of the forest."*
>
> *Ali F. Oeus*

From birth through early developmental stages, there is a debate as to what is more important, nature versus nurture. Nurture being elected human behavior, versus nature, the exposome-genome entanglement. Within this context, nature is a deterministic concept, while nurture is modifiable and reciprocally modifies the effects of nature. Traits and developmental stages are programmed into the code, but what is expressed is heavily dependent on the environment. While both concepts of nature and nurture are important, TVST leans towards the environment being the more potent and important contributor to who humans are—the result of the exposome's engagement with the genome—as opposed to human's behavioral choices. [354] Life's course through the spectrum of threat versus safety causes genetic and epigenetic modifications to the genome's expression that modify the developmental processes across a lifetime. Yet, some of life's course is determined by the observations, choices, and behaviors humans employ across a lifetime which in a bidirectional fashion push back at the epigenome and exposome.

Historically, aging and death are not seen in the same way as birth and development, but maybe they should be. Aging seems to be a programmed biologic necessity, a developmental process.

Aging is characterized by gross phenotypic changes and alterations in cells, tissues, organs, and organisms. Various hypotheses attribute aging to DNA damage, progression of cellular clocks, free radical damage, and cellular distress responses. There is a narrative that damage to molecules, cells organelles, and the extracellular matrix seem to be involved in aging. At the molecular level, genomic, epigenomic, and proteomic damage are postulated to contribute to aging-related phenomena such as stem cell exhaustion, mitochondrial dysfunction, energy metabolism failure, altered cell signaling, and, ultimately, impairments of biological functions. Conclusions include that aging is a multifaceted phenomenon influenced by various biological processes. The epigenome and proteome have garnered most of the attention in aging, but recent investigations of the transcriptome have demonstrated that morphologic changes to the length of the genes themselves is also associated with aging.

Much of aging looks just like a cumulative and compounded chronic TR. Threat, TS, TC, and the associated TR i.e., threat associated distress contribute to the alteration of genes, telomeres, chromosomes, and aging.[355] Some of these changes reverse with rest and recovery—safety. The shortening of telomeres is correlated with aging, if not an accelerant of aging.[356,357] In addition, an elevated TL creates clear epigenetic changes that correlate with an increased rate of aging. This creates an epigenetic clock that can be used to define biologic age, as opposed to chronologic age, that predicts death. [358,359,360]

Epigenetic Age = Time + TL

(Epigenetic age may be a misnomer in some sense as a component is the TL. For humans time is not reversible but the TL, to some extent, is reversible. The TL tends to increase with age. However, it is a reflection of accumulated distress and compounded threat more than a reflection of age and time—or time left. In addition, "epigenetic age" or the determinants of epigenetic age fluctuate or oscillate throughout the day; it is neither stable nor linear in progression. It seems advisable not to focus on "epigenetic age" but to focus on epigenetic threat coding and what can be modulated.)

The epigenetic patterns and aging that are associated with increasing threat, TS, TC, and an associated TR are sometimes referred to as "inflammaging."[361] It is an interesting spin on language that again fails to see the complexity of a TR and minimizes the diffuse and systemic nature of a TR and its metabolic componentry.

Aged mitochondria, cells, tissues, organs, and humans look progressively like T2 phenotypes, with lethargy and hypometabolism; impaired oxidative phosphorylation and anti-oxidant formation; increased reactive oxygen species and oxidative enzyme production; impaired recycling, regeneration, and reproduction; increased catabolism and inflammation; impaired multiple tissue and organ function; stiffness and stooped posture; declining activity, mood and cognition; generalized degeneration and senescence.[362] It is interesting to note that the aged and those relatively young sufferers of schizophrenia have similar brain changes and associated declining cognitive flexibility and loss of neuroplasticity—time versus TL?

Threat, TS, TC, and an associated TR induce senescence in cells. Senescent cells produce large amounts of TC to increase TS and a TR. This can become a feedforward system, resulting in a steady physiologic decline. The TC can initially signal to immune cells to deconstruct the senescent cell. But too much threat signaling induces the T2 phenotype with "immune-senescence," "immune-paralysis," or "immune-exhaustion" and the impaired ability of the immune cells to deconstruct the senescent cells.[363] There is a critical value, perhaps 2-3%, of senescent cells to total cells from which there is no return. From this point, disease, disability, pain, suffering, and aging accelerate.[364] Death follows.

The number one risk factor for all chronic diseases is age—80% of all people over sixty-five suffer from at least one chronic condition. After sixty-five years old the risk for Alzheimer's disease doubles every five years. By eighty-five years, one third of all people suffer from Alzheimer's disease, while one third of men, and one fourth of women will suffer from some form of cancer.

To put things into further context, tortoises have extremely long lives. Some live up to 200 years. Within the right environment, they can markedly slow cell senescence. Perhaps having a hard shell for armor and defense lessens the need for an aggressive oxidative, inflammatory, and catabolic TR. An absence of such threat physiology, particularly in a chronic capacity, could allow for longevity.

Similarly, whales and elephants have long lives. Could their size be enough of a deterrent and defense that they also have a relatively quiescent TR? Notably, all these species have very low rates of cancer, also suggesting a less aggressive TR. And within cancer cells, there has been a demonstration of genotypic plasticity and phenotypic change, instead of immunologic destruction, to resolve cancer growth in these species.[365] It seems worth studying tortoises, whales, and elephants.

What could humans use as an armor or a defense for this kind of protection? A will to keep each other safe?[366,367]

All this begs the question as to how much of aging can be attributed to nurture, elected human behavior, versus nature, the exposome-genome entanglement? And what is the potential longevity of the genetic code of safety? Those answers are not entirely clear. However, the failure to protect and nurture across a lifetime certainly effects overall health and longevity.[368]

Perhaps, a more interesting question is: how can compounded genetic and epigenetic threat coding be deconstructed? How can the transcription and translation of the code that results in senescence, degeneration, illness, disease, disability, pain, and suffering be turned down or off?

A hallmark of eukaryotic aging is a loss of epigenetic information, a process that can be reversed. A recent study has shown that the ectopic induction of certain transcription factors in mammals can restore youthful DNA methylation patterns, transcription profiles, and tissue function, without erasing cellular identity. This is a process of active DNA demethylation without altering the genome. Pharmaceuticals for epigenetic stripping are right on the horizon. Could this be a mechanism by which to decode the threat code, to initiate safety recoding and healing?[369,370] Could this be a reset that might alleviate the spectrum of threat related illness, disease, disability, pain, and suffering?

In addition, how can epigenetic safety coding be constructed and compounded? How can the transcription and translation of that code for vitality, growth, repair, regeneration, reproduction, longevity, health, and wellness be activated? What individual and societal structures can be constructed to armor and protect the fragile human mind and body—protect the system? In addition, what are the pharmacologic opportunities to signal safety?

Humans diverged phylogenetically from the tortoise 300 million years ago. Do humans still have the genetic potential to live 150 years after all this time? Probably, with the right armor. People who age more slowly than their chronologic peers have been noted to have:

- Higher cortical brain volumes
- Better episodic memory
- More education
- A musical background
- Preserved muscle mass (especially fast-twitch fibers)
- Better agility and balance
- Better social relationships
- Better sleep
- Less anxiety and depression
- Better cholinergic-parasympathetic function

Interestingly, where they do not differ is in the amount of exercise they get, their APOE status, and their beta amyloid or tau protein accumulation. APOE4 and tau protein are some of the classic markers for dementias and other neurodegenerative diseases—are they truly meaningful?[371]

Memory games, high intensity training, sociality, sleep, and happiness can all slow aging to some degree. However, all the correlates with the phenotype of slower aging are also correlates with being safer and the physiology of safety. Again, TVST postulates that the key is to reduce threat and increase safety within people's lives and the world to afford these better outcomes in illness, disease, disability, pain and suffering, and slow aging.

The art of aging is moving through the final developmental stages to include letting go of the false-self, connecting with the other (teaching, mentoring, experiencing), and being a safe vector for, not an authority on, the knowledge and wisdom of life. Knowing when you are no longer the show and when it is time to let go of those illusions is not only protective and propagative for the species, but also helps to prepare for the final dissolution—death. If the skill of life is finding safety, the art of life is letting go.

There is a time to lead, a time to follow, and a time to get out of the way. Flexible faith and willing acceptance can come with age if this aging is in a place of safety. Aging in threat is much more likely to be associated with rigid belief and defiant resistance. In this regard the aging process is contextual to threat versus safety.

Finally, consider aging and dying under the light of apoptotic programmed cell death. Is the final dissolution a necessary social act of sacrifice for the greater good, and the defense, protection, and propagation of the human genetic code—the survival of the species?

Are humans already living too long?

Fascio-musculo-skeletology

> *"It is health that is real wealth and not pieces of gold and silver."*
>
> *Mahatma Gandi*

Even fascio-musculo-skeletal health falls into the threat versus safety paradigm. Osteoporosis is a catabolic disease influenced by threat, TS, TC, and an associated TR.[372] Arthropathies/itis and tendinopathies/itis (degeneration, inflammation, and/or pain in FMS structures) are both catabolic and inflammatory, with associated tissue loss and pain.[373] Arguably, gums, tendons, cartilage, spinal discs, and even bone, are targets in a TR because of their rich stores and sources of fuel, having high levels of proteoglycans (stored sugar) for the metabolic needs of fight and flight.

Large proteoglycans are essential for gum, tendon, cartilage, and spinal disc hydration, flexibility, and load absorption. Mobilization of these proteoglycans results in stiffness, rigidity, dysplasia, and degeneration. Proteoglycans through sulfated glycosaminoglycans regulate cell signaling, tissue development, and growth and regeneration. TC induce mobilization and catabolism of proteoglycans.

Osteoblasts decrease activity, and osteoclasts increase activity with TS. Bone is broken down when in threat. Calcium, an inflammatory molecule that enhances a TR from its actions within the blood vessels to its enhancement of threat signaling within the nervous network, is mobilized from bone, along with fuel substrates within a TR. Distress, inflammation, bone catabolism, and calcium mobilization are also significant risks for kidney stone formation. Air pollution, notably, decreases bone quality and increases the risk for bone fractures, but so do other threats.[374] Interestingly, the antithreat monoclonal antibody used to treat osteoporosis also has benefits in controlling obesity and diabetes.

Early onset degenerative arthritis has doubled since 1990. This trend parallels the obesity epidemic, but obesity should be considered a correlation, not a causation of degenerative arthritis, as they both arise from a common root and soup of threat. To reverse degenerative bone, joint, disc, tendon, and gum disease, it is important to not only block TS, but to increase SS. Many medications (nonsteroidal anti-inflammatories drugs [NSAIDs], glucocorticosteroids drugs [GCSDs], and IL6 inhibitors) block this transition to healing. Other newer medications (monoclonal antibodies, small molecule inhibitors) only block TS without improving SS.

In addition to the phenotypic changes associated with threat in the osteocytes, chondrocytes, synoviocytes, and fibrocytes of bone, cartilage, joints, and tendons, the phenotypic changes in the fibrocytes of the fascia have direct implications in the symptoms of myofascial pain (T1 phenotype) and fibromyalgia (T2 phenotype).[375] Even the immune cells' threat associated phenotypic and functional changes play large roles in fascial pain syndromes.[376] Temporomandibular joint syndrome is a myofascial physiologic condition, not a structural problem, that can fall into a T1 or T2 phenotype. Frozen shoulder syndrome is a T2 phenomenon that transitions from a hot inflammatory "capsulitis" phase to a cold fibrotic "capsulopathy" phase—T1 to T2.

You don't get the butt you want by sitting on it.

Indeed, fascio-musculo-skeletal disease cannot be separated from the greater system and therefore seldom exists in isolation. Stress in threat (distress) increases TS, TC, inflammation, catabolism, and degeneration. Stress in safety (eustress), such as exercise, increases SS, SC, healing, anabolism, and regeneration.[377] Distress is stress when in threat. Eustress is stress when in safety. Distress makes us weaker. Eustress, such as play, labor, and exercise, if interspaced with rest and recovery makes us stronger. What doesn't kill us doesn't necessarily make us stronger—stress is relative to threat versus safety. Chronic distress does weaken and kill us.

Contemplate this: threat increases rigidity of cell membranes, tissues, organs, social contracts, cognitive constructs, symbolic narratives, and held beliefs. Safety increases flexibility in all those areas. Rigidity leads to brittleness and an increased risk for injury in physical and other forms.

Threat-associated white fat accumulation is proTR as white fat acts as a paracrine and endocrine organ to promote inflammation and insulin resistance through the extracellular release of TC, *adipokines*. By contrast, muscle can act as a proSR paracrine and endocrine organ through the extracellular release of SCs, known as *myokines*. Myokines are cytokines that are produced, expressed, and released by muscle cells. Myokines may balance out and counteract the effects of adipokines. There is an increased release of these muscle SC with exercise, "*exerkines*," that leads to both local and systemic benefits.

Myokines mediate the protective and anabolic effects of

muscular exercise and mitigate diseases associated with a physically inactive lifestyle. They are fundamentally anabolic and promote the resolution of inflammation. Myokines cross the blood brain barrier to exert protective and anabolic effects on critical brain structures, particularly the neocortex. Myokines are protective from anxiety, depression, cognitive decline, and dementias. Safe activity promotes the expression of these healing and health related cytokines.

It is worth mentioning here that the modern concept of exercise would be foreign to human ancient ancestors. Traditionally, exploring, hunting, foraging, gathering, laboring, and playing provided the activity the system required to be healthy and happy. Labor is as integral to the species as sex. The value of labor has been marginalized in the modern world. It should be included as a determinant of health and incorporated into treatment algorithms—play too. Notably, labor is work, but not all work is labor. Labor is movement, exertion, and creation. White-collar jobs do not afford the physiologic benefits of labor. The modern concept of exercise should be secondary to exploration, labor, and play, a backup when exploration, labor, and play are inaccessible.

"Genius is 1% inspiration and
99% perspirations."

Thomas Edison

Truly, the growth and health of the brain is more dependent on physical exercise than cognitive exercise. In fact, excessive cognitive strain can make us sicker, and weaken the brain. Chronic cognitive loading increases TCs and threat phenotypes with increasing irritability, anxiety, fatigue, muscle tension, soft tissue pain, and paradoxically poorer cognitive function. The desire for white collar work is a construct, narrative, and belief within our minds whereas the drive to labor and create arises within our DNA.

The threat-induced changes in monoamine profiles associated with shifting phenotypes not only have effects on the brain, mood, behavior, cognition, and even immune cell function, but they also directly affect reflexes, muscle spindles, tone, and tension,[378] and facial expression and body posture. Thus, they play a part in myofascial disorders, chronic pain, along with fascio-musculo-skeletal and spine deformities.

In general, there is a progression from being upright in both S1 and T1 phenotypes and being more protected and kyphotic (a curving of the spine that culminates in a bowing or rounding of the upper back), if not fetal, in both S2 and T2 phenotypes. Kyphotic and scoliotic deformities may be mediated by threat physiology. Most notably dopamine, serotonin, and melatonin production and transmission are impaired by TCs. Dopamine depletion is associated with kyphotic postures. Serotonin depletion has been associated with scoliosis in some species.[379] Implanted metal hardware and boney fusions to straighten a degenerative and/or deformed spine are poor substitutes for the provision of safety and a straight and healthy spine.

Back pain is the leading cause of disability worldwide. Most neck and back pain are neither from an injury nor a postural or structural problem. It is a physiological problem that will not be cured by pills or procedures, but by safety.[380,381] Treating a physiological problem with structural interventions seldom results in benefit, while it places patients at significant risk for harm, including complications, acceleration of disability, and even death.[382]

Conflict, including internal emotional, behavioral, and cognitive dissonance, has been associated with increases in neck and back pain.[383] The vast majority of neck and back pain is not from injury or structure but is physiologic and soft tissue in nature. Moral injury from a sense of insult, defeat, betrayal, shame, or guilt has been associated with increases in soft tissue and joint pain.[384] For the general population, and especially clinicians, it is very important to recognize that connective tissues can hurt without a structural injury or precise anatomical source.

Over the last thirty years, spinal procedures have grown exponentially, while outcomes for neck and back pain have declined. Greater than 60% of all spinal procedures are likely unnecessary. Injections, ablations, and fusions are conducted every day across the United States for physiologic connective tissue pain—threat disorders. Low-value spine imaging leads to over utilization of low-value spine procedures that do not help and can harm.

Patients undergo many structural procedures for nonstructural pain. These include fusions from the head to the pelvis on normal spines that leave patients immobile, deformed, and at risk for real structural problems, paralysis, and more surgery.

Even relatively simple pain procedures can go awry and lead to hemorrhages, infections, and permanent paralysis without affording a cure. The source of the pain is not an abnormal structure. Its source is not local. It is a systemic process from the root and soup of chronic threat. The source of much chronic pain and suffering is chronic threat.

Pain, stiffness, stooping, and wearing out arise from the root and soup of threat and are more than just the culmination of aging. These changes are very much related to the mitochondrial, cellular, tissue, and postural phenotypic changes associated with threat.

"The magic potion is knowing there is no magic potion ...there is only the reality that surrounds us."

Ali F. Oeus

Consider that if TC induce a variety of catabolic and degenerative changes in the body, then the SCs do essentially the opposite. As noted previously the TC tumor necrosis factor alpha (TNFa) was initially identified as a hormone, cachectin, due to its causing of "wasting." All the TC cause wasting. Loss of muscle mass, sarcopenia, has been noted as a risk factor for Alzheimer's disease. The screening for loss of strength has been called for as a method to screen for the risk of dementia. TVST would postulate that sarcopenia is correlative, not a causative, nor a risk factor for Alzheimer's disease. The two come from the same root and soup—threat, threat phenotype, threat physiology.

Therefore, loss of strength is a good screen for chronic threat **and all** threat-associated diseases.

Exercise, in those who have a tolerance for it, releases safety cytokines to stimulate the building of brain, muscle, bone, and other tissues. Exercise is therefore a good tool for the prevention of Alzheimer's disease, as well as many other threat-associated diseases. However, in Post Covid-19 syndrome, chronic fatigue syndrome, and myalgic-encephalomyelitis syndrome (T2 phenotypes), excessive exercise can make physiology and symptoms worse and is associated with post exertional malaise. In deep threat phenotypes, the cells and their mitochondria lack the directive and ability to promote SC and healing. Excess or additional stresses in threat states can be a form of distress and can perpetuate the TR and deepen T2 physiology.

Post-exertional malaise or exhaustion is common in deep T2 phenotypes. In these patients, and really all patients, safety must be primary in a treatment algorithm. Exercise becomes a secondary intervention for healing once the phenotype and physiology are shifts and the benefits of eustress are possible. Safety is preventative, healing, curative, and primary for many people with chronic diseases.

Exercise is important, especially in the forms of play and labor, for human beings. The cultural revolution of the 60s and 70s included not only a focus on personal freedoms and civil rights, but new interests in concepts of self-care, nutrition, and exercise. While the social revolution has sputtered as of late, self-care, nutrition, and exercise continue to preoccupy us to this day. It is noteworthy that nutrition and exercise may account for only 10-20% of the determinants of biologic aging and health.[385] There is also a diminishing marginal return with exercise in terms of health and longevity—more is not always better. TVST postulates the vast majority of the remaining 80-90% of the determinants of aging can be attributed to threat versus safety within the world. Will the decades to come be the start of a new focus, a new cultural revolution, to move beyond the individual towards societal and global care and safety?

TBD

Sleep

"Sleep that knits up the ravell'd sleave of care,
The death of each day's life, sore labour's bath,
Balm of hurt minds, great nature's second course,
Chief nourisher in life's feast"

Macbeth – William Shakespeare

It is worth mentioning the function of sleep. Quality sleep is found when in a S2 phenotype. It is a time of healing—physically, emotionally, and mentally. It is a time out of awareness for the system to recode, to reboot, to grow and regenerate cells, tissues, and organs, and to prune information and consolidate memories. It is a time for the body and brain to cleanse itself as well. It is in the safety of sleep that humans find resilience, health, and wellness.

Threat, TS, TC, and the associated TR maintain vigilance and destroy safe restorative sleep. Nocturnal movement disorders, bruxism, nightmares/terrors, and early awakening are all threat based and interfere with good sleep. Obstructive sleep apnea is frequently the result of excessive white fat deposition in the soft tissues around the upper airway, but is also associated with PTSD exclusive of weight. Obstructive sleep apnea, along with obesity, are threat-based problems that increase TS and TC to exacerbate the TR.[386] Rapid eye movement sleep disorders are associated with threat-related disorders and diseases, and highly correlate with neurodegenerative disease.

Threat, TS, TC, and the associated TR also loosen the blood brain barrier and impair the glymphatic cleansing of waste products that should occur when sleeping. This can lead to a toxic congestion of the brain with deterioration in cellular brain functions. Sleep restores this balance, but it requires safety. It is worth noting that glymphatic flow is actually reduced during sleep, as are many physiologic functions. It is safety physiology in both awake and sleep states that is important to blood brain barrier integrity, and the cleansing, and recycling functions of the brain.

The biologic molecular clock is largely regulated by daylight, or lack thereof. Both daylight exposure and darkness exposure are important to health and vitality. Human cells—especially retinal cells and skin cells, but all cells to some degree—sense light and change in light. In this respect, humans can be tightly regulated by and integrated with the natural world's sunrises and sunsets. The modern world has destroyed this connection and discombobulated human cells. Excessive light exposure disrupts the connection to the rhythms of the natural world and the mechanics of the biological molecular clock. This is especially true of the blue light produced by technological devices, particularly excessive exposure to this light in the afternoon and evening hours. Excessive blue light exposure at the wrong time is an environmental pollutant of sorts and acts as a threat to us.[387]

Chronic loss of sleep is associated elevated TS and TC within a TR and all of the other sequalae of a TR:

- Oxidative stress
- Mitochondrial and cellular changes
- Immune and metabolic perturbations.
- Altered autonomic functions (temperature, heart rate, blood pressure)
- Elevated distress hormones

This culminates in an increased risk of obesity, type 2 diabetes, metabolic syndrome, cardiovascular disease, cancer, cognitive impairment, neurodegeneration, epilepsy, psychiatric disorders, threat bias, loss of empathy, increased prejudice, and asocial and antisocial behaviors.[388,389,390] Nocturnal workers have a higher incidence of all these afflictions.[391]

Air pollution can activate a TR and reduce the quality of sleep. In fact, prenatal, perinatal, and postnatal air pollution exposure decreases the quality of sleep in children years later.[392] This is consistent with TVST and the inclusion of generational, historical, and ecological threats within the total TL. A comprehensive review of threat is critical to understanding the phenotype of a human presenting before us within clinical health care.

Lack of sleep is toxic to the system. Threat interferes with sleep. Lack of sleep increases TS and TC, thus interfering further with sleep—a downward spiral for which safety is the solution. Every threat-related diagnosis can be caused or exacerbated by lack of sleep. Lack of sleep can be caused or exacerbated by every threat-related diagnosis. It is a bidirectional rut that gets dug deeper and deeper when humans are stuck in threat.

Thus, the provision of safety must include the construct that sleep is good, sleep is necessary, and sleep is not to be cheated. There is no shame in getting a good night's sleep. There is ample evidence that this is all true, but humans still suffer from the cultural construct that being able to survive on little sleep is a sign of strength—it is not. It is a sign of foolishness. It breeds illness, disease, disability, pain, suffering and weakness.

Women's Health

> *"If you can dance and be free and not embarrassed, you can rule the world."*
>
> *Amy Poehler*

TVST postulates that it is the exposome that signals the system to adjust and express the genome selectively and strategically. The spectrum of this expression composes the phenome.

TVST sees the exposome and the genome as collaborative parts in the expression of the phenome. In most cases, the exposome is the dominant factor in illness and disease, as it tends to dictate the expression of the genome, as opposed to human choice and behavior pushing back against nature to nurture.

In addition, TVST postulates that the genome for SS is always present and retrievable if the exposome supports safety. TVST suggests that even in dysfunction, degeneration, and disease, humans are still whole—not broken—at a genetic level.

Admittedly, there is some genetic variation within individuals and populations, as well as from the plasticity of the genome and epigenome, that determines some differences in an expressed TR. Generally, these genetic differences account for the inherited traits, but not the variability or mutations seen in illness and disease. They are not the source of most illness and disease. Threat is the source. These genetic differences provide for the genetic variability of a response to threat.

Having said all that, there is one large and striking genetic difference in humans that deserves special attention. Humans typically come in two fundamental genetic forms—female and male—with physical, emotional, behavioral, social, mental, and spiritual implications for illness and disease versus wellness and health. In general, females across the globe suffer a lack of recognition and representation. They endure more oppression and repression throughout their lives, including in science and medicine. Girls' and women's burdens of threat are higher in almost all cultures. Additionally, human females experience many more programmed inflammatory events within a lifetime than their male counterparts. Not surprisingly, this all has a direct effect on their health, happiness, and quality of life.

Women with a history of domestic violence and abuse are at greater risk for asthma, atopic dermatitis, and allergic rhinoconjunctivitis. Women who experience childhood trauma have an increased risk for smaller cortical brain volumes and both primary and secondary psychological disorders.[393] Cumulative and compounded TL, with associated phenotypic and physiologic responses to trauma, lead to changes in neuro-immuno-regulation, increases in TS and TC, as well as other threat-associated neuropeptides. IgE also increases as immune cells change to threat phenotypes.[394] These same threat phenotypes, years later, are associated with an increased risk for cancer.[395]

For millennia, threat phenotypes in women were felt to have their origins in the uterus. Hysteria was a formal psychiatric diagnosis used to describe this condition. The Greeks described the problem as a wandering uterus. Others felt demonic possession was the problem. Within the past two hundred years hysterectomies were performed to cure the phenomenon.

This all sounds ridiculous to us today. But to put this into perspective, the real question is which dogma of today will be ridiculous tomorrow? Starvation for weight loss? Spine surgery for back pain? Stenting for heart disease? Chemotherapy for cancer? What about our current understanding and treatments for endometriosis? This disturbing misunderstanding and mistreatment of women around the diagnosis of hysteria became symbolic for the women's suffrage movement at the beginning of the 20th century. Progress has been made, but some darkness and blindness persist. TVST hopes to illuminate and bring to focus persistent misconceptions.

"The body keeps the score."

Besel Van der Kolk

Females go through many more programmed inflammatory and catabolic events in a lifetime than males. Menarche, menses, pregnancy, labor, and menopause are all associated with elevated TS and TC, irrespective of TL. This very likely predisposes females to many inflammatory and catabolic illnesses and diseases. Four out of five auto-immuno-inflammo-catabolic diseased patients are female. Menstrual irregularities and ovulatory disorders are associated with cardiometabolic disorders such as obesity, diabetes, dyslipidemia, hypertension, insulin resistance, metabolic syndrome, polycystic ovary syndrome, vascular heart disease, valvular heart disease, arrhythmias, transient ischemic attacks, strokes, coagulation disorders, deep vein thromboses, and pulmonary emboli.

Each of these events is associated with phenotypic changes. The ebb and flow of cytokines influences not just hormone levels and reproductive physiology but also systemic physiology, reproductive anatomy, and systemic anatomy including the brain. Phenotype changes are associated with emotional, behavioral, social, mental, and structural changes. Observed with MRI imaging, cortical and subcortical brain structure and function shift throughout the menstrual cycle, and brain circuitry conduction is faster in the less inflammatory phases of the cycle. A stale endometrial lining presents as a threat with increased TS, TC, and an associated TR. This is a normal acute threat that the body deals with through sloughing and expulsion. Cyclical yet constant change, thus infinite selves, is the flow of life—so much for concepts of stasis.

Notably, a minority of fertilized human eggs result in a full-term pregnancy and a live baby, and the majority of these lost cells would not be classified as frozen embryos, miscarriages, or abortions. An increased TL further reduces the likelihood of a successful fertilization of an ovum, endometrial implantation, healthy pregnancy, and an uncomplicated live birth. Threat blocks impregnation and causes miscarriages at a much higher rate than contraception or abortion. Infertility isn't a disease, it is a phenotype of threat. Removal of women's status and rights amplifies threat signaling creating more infertility, disease, disability, pain, and suffering.

Infant preterm births, morbidity, and mortality, as well as maternal complications and worse outcomes from pregnancy, labor, and delivery, are associated with increased threat, TS, TC and an associated TR.[396,397] These threats include higher exposures to ecological threats such as climate warming, air pollution, and chemicals (phthalates) found in plastics[398], but also all other ecological, physical, emotional, social, mental, financial, cultural, societal, and spiritual threats—the total TL.

The Maternal Vulnerability Index (MVI) is an index that quantifies maternal vulnerability to adverse health outcomes associated with preterm births that reflect the physical, social, and health care landscapes of the mother. It is organized into six themes:

- Reproductive health care
- Physical health
- Mental health and substance abuse
- General health care
- Socioeconomic determinants
- Physical environment

A very high MVI is associated with increased preterm birth.[399] From 2014 to 2022, preterm births increased 42%—a trajectory from increasing threat.

Maternal threat is associated with increased miscarriages, stillbirths, low birth weights, and a child's illness and disease burden throughout a lifetime.[400] This is commonly attributed to socioeconomic inequities (classic social determinants of health), but the TL runs much deeper and wider than these inequities. All threats are detrimental to a mother and child.

Additionally, complications in pregnancy are predictive of later life threat-related disease in mothers, as they frequently arise from the same root and the same physiologic soup of threat.[401] Epigenetic research demonstrates that threat to a mother and baby in-utero and threat to a child in early life negatively affects the child and their health throughout a lifetime.

In 1998, a severe snowstorm in Montreal knocked out electricity for weeks. Profiles of women pregnant during this time and their children demonstrated the amount of time they were without electricity adversely correlated to the child's intellect, body mass index, insulin secretion, risk for diabetes and immune functions. The TR is transferable across the placenta to influence later development. These changes are felt to be epigenetic in nature, thus potentially reversible with the decoding of threat and the recoding of safety.

Maternal health and happiness must be a priority for the health of the species. The spectrum of women's health issues includes:

- Subordination
- Disenfranchisement
- Discrimination
- Abuse
- Rape
- Menarche

- Premenstrual syndrome
- Pregnancy
- Preeclampsia
- Eclampsia
- Labor
- Delivery
- Toxic shock syndrome
- Post-partum depression
- Endometriosis/itis
- Adenomyosis/itis
- Fibroids
- Polycystic ovary syndrome
- Idiopathic intracranial hypertension
- Interstitial cystitis
- Pelvic pain
- Lichen sclerosus
- Menopause
- Hot flashes
- Osteoporosis
- Eating disorders
- Obesity
- Diabetes
- Metabolic syndrome
- Stroke
- Heart disease
- Cancer
- Anxiety
- Depression
- Dementia
- Chronic pain
- Auto-immuno-inflammo-catabolic diseases

In addition, women have a higher incidence of postural orthostatic tachycardia syndrome, myalgic-encephalomyelitis/

chronic fatigue syndrome, and post Covid-19 syndrome than men do.

All these events, conditions, and disorders are associated with increased TS, TC, and associated TR, and a high TL makes all these issues worse.

Menopause deserves some contemplation. Menopause is when menstrual periods stop, and women are no longer able to bear children. It is also defined by a decrease in steroid sex hormones. But more specifically, menopause is associated with the senescence of ovarian cells, particularly the oocytes or eggs. This deterioration is associated with a gradual increase in an inflammatory, oxidizing, fibrosing, and catabolizing microenvironment that hits a threshold with systemic effects from an exponential jump in TC. Symptoms include:

- Hot flashes
- Flushing
- Night sweats
- Arthralgias
- Skin deterioration
- Insomnia
- Mood changes
- Cognitive decline
- Increased abdominal fat
- Elevation of lower-density lipoproteins and triglycerides

Risk for obesity, diabetes mellitus, cardiovascular events, osteoporosis, dementia, and mental illness rise after menopause.

This picture looks much like a TC event, perhaps more than a hormonal event. Indeed, TC elevate with age, but there is a larger spike with menopause. Threat, TS, TC, and an associated TR all result in a decline in sex hormone production. Is this decline a correlate and/or cause of symptoms? In turn, this becomes a feed-forward process. A decline in sex hormones can further exacerbate TS and TC levels since there is bidirectional signaling between steroid sex hormones and cytokines.

Consider that aging involves a cumulative and compounded TL. Could it be that when TC reach a critical level, they cause a phenotypic shift that precipitates menopause? Menopause may be an evolutionary adaptation to prevent women from bearing children that they may not be able to raise due to their advanced age, or it may simply be cumulative and compounded threat. Perhaps both.[402]

Certainly, distress affects menses and fertility. Menstrual periods become irregular with distress. An increased TL is associated with premenstrual syndrome, dysmenorrhea, and menorrhagia. Fertility declines with distress.

Menopause is precipitated and accelerated with distress. Proper management of menopause may involve managing the TL, steroid sex hormones, and TC and SC. [403,401,402,403,404,405,406,407,408,409,410,411,412] Osteoporosis is already managed with a monoclonal antibody threat signaling inhibitor. What else could or should be managed this way, and what should be managed with a safety facilitator?

Given the expanse and consequences of women's unique health issues, it becomes an imperative to view these issues through the lens of TVST and to develop comprehensive sociopsychobiological health care models for the mothers and daughters of the world.

When abandoning a ship, it used to be women and children went first, old men and the captain went last. Societal and cultural constructs should prioritize women and children first. The white male dominance of medicine in the Western world has failed to do this within research, clinical care, public health, and sociocultural care.

Men's Health

> *"The mass of men lead lives of quiet desperation."*
>
> *Henry David Thoreau*

Women's health issues may not exceed men's health issues, but they have been relatively under studied and marginalized more so than men's issues. Men do die at a younger age than women (U.S. average lifespan for men is consistently over time about six years less than for women). There are a variety of reasons for this disparity, including taking higher risks, more dangerous jobs, closer relationships with guns, lack of close social relationships, higher rates of cardiovascular disease, and being less engaged with their healthcare and the health care system.

TVST can be used to dig into each of those issues, but most of those issues have been addressed within other sections. However, it would be remiss not to include some commentary on men and their issues. Inspection of what threat does to sexual and reproductive functions is an avenue to explore in both men and women, but here the focus will be weighted towards men.

As noted, the spectrum of a TR shifts priorities and mechanisms within physiology. In general, reproduction is not a physiologic priority for men or women when in threat. This is reflected in peptide and hormone levels with a reduction in SC and sex hormones and an increase in TC and distress hormones when in threat. This can lead to sexual and reproductive dysfunction in both men and women. The spectrum of dysfunction includes impaired fertility, reduction in sex drive, and reduction in sexual performance for both sexes. It is technically incorrect to term this dysfunction, as the system is doing exactly what it is supposed to do, given the TL that is being interpreted and transcribed into physiology.

For men there are multiple sexual phenotypes of threat. Notably, bacteria (and mitochondria) in threat undergo fission. Reproduction is a defense strategy to keep the genetic code going by rapidly dividing, multiplying, and scattering. Complex multi-cellular and differentiated-cellular organisms do not have the option of fission. The reasons are that reproduction is sexual, dependent on another, and the organism is dependent on all its different cells, tissues, and organs for survival—mitochondria and cells can split in half, but humans can't survive such an event. Cancer cells separate from the whole and do rapidly divide and replicate to a poor outcome for both the whole and the cancer cells. However, humans do have some other mechanisms for saving the code that can be employed in threat states.

Characteristic of a male in early threat states, T1 phenotypes, is sexual dysfunction in the form of premature ejaculation. It does make some biologic sense to try and quickly spread the code and then return to a defensive position when in threat. In T1 phenotypes, infidelity is more common—a short-term strategy to spread the code to more partners, improving the chances of the code making it to the next generation.

Threat, TS, TC, and an associated TR influence endocrinologic hormone and neurologic transmission and autonomic networks. The sympathetic nervous system is very involved in this process. The parasympathetic nervous system is involved in safety physiology, connection, bonding, arousal, and foreplay and is relatively inactive in threat states. It is the sympathetic nervous system that engages in the performance and orgasmic phases of sex. Beta adrenergic signaling increases heart rate and blood flow to the muscles, and if sexually aroused, the genitals (also, it conveniently decreases bowel and bladder activity conveniently). As excitement escalates, alpha1 adrenergic signals close off the bladder neck and facilitate orgasm and ejaculation. [413] The alpha1 adrenergic signals are also vasoconstrictive, resulting in detumescence and a refractory state for sexual activity. Ligand concentrations and receptor affinities play a large role in this progression. High levels of adrenergic ligands will not only predispose to premature ejaculation, but to sustained refractory periods and erectile dysfunction.

Notably, the T1c phenotype is an adrenergic dominant state, exhibiting more retreat, hypervigilance, and compulsive behaviors than T1b. The latter is a dopaminergic and adrenergic state, exhibiting more approach, aggressive, and impulsive behaviors. The T1b state provides the urges of mind and body whereas the T1c state produces the urgency of mind, body… and bladder. Therefore, arguably due to the hyperadrenergia of T1c, anxiety and sexual dysfunction would be expected to be higher in this state.

One can see that in a T1 phenotype, sexual activity is still possible, as it is very dependent on the activity of the sympathetic nervous system. It should also be noted that the sympathetic nervous system can be active in both T1 and S1 phenotypes. Therefore, it should not be misconstrued as or conflated with the fight or flight system. It is contextual and can be active in both aggression and love.

Even though the system is deprioritizing sexual functions in T1 threat states, sexual function is altered, but still possible. However, T1 lacks the intimacy of sexual engagement seen within a S1 state. This then raises the issue that the T1b phenotype is characterized by irritability, anger, rage, explosivity, impulsivity, and antisocial behaviors. The T1b phenotype is physiologically a setup for sexual assault and rape.

Once into a T2 phenotype, the picture is very different. In this state desire, performance, and fertility are all impaired. Here, both parasympathetic and sympathetic functions are decreased. Therefore, not only are connection, bonding, arousal, and foreplay disengaged, but sexual performance is more markedly impaired as well. This is characterized by low drive, poor arousal, erectile dysfunction, and delayed or absent orgasms. T2 states are also characterized by relative isolation and asocial behaviors, more so than the antisocial behaviors seen in T1 states. In addition, these states are characterized by ruminations and obsessions that can escalate into maladaptive sexual ruminations, obsessions, and perversions. Narratives and beliefs born and rehearsed in T2 states could thus potentially facilitate T1c compulsive and T1b impulsive antisocial and predatory behaviors.

So much of men's physical, emotional, behavioral, social, mental, and spiritual health issues are directly related to the total TL. Therefore, when diagnosing and treating sexual dysfunction prior to seeking a multitude of laboratories, prescribing a plethora of pills, or performing a variety of cardiovascular, gastrointestinal, or genitourinary procedures, it can only make sense to start addressing these issues at the root—chronic threat. It is important to remember that psychology is physiology, expressed through feelings, behaviors, and thoughts at the level of the big brain, yet, a man's little brain is not separate—there is no mind-body duality.

"Winning isn't everything but
wanting to win is."

Vince Lombardi

Within the context of men's health, it is worth noting that winning is an act of aggression—winning, domination, teasing, and bullying have their foundations in threat. Yet they feel good, they are associated with a sensation of pleasure, they have higher dopamine levels and lower inflammatory markers than other threat states, and these behaviors can prevent slipping into these deeper threat states. However, they must be repeated over and over again to sustain this high.[414] These behaviors are addicting. Additionally, dopamine transmission is not solely about movement, mobilization, motivation, reward, and pleasure. Dopamine transmission is also associated with approach, aggression, and early defense programs—fight. Pleasure, aggression, and domination are thus chemically closely approximated. Context of threat versus safety determines the expression of dopamine states.

Conversely, connection, cooperation, and adaptation have their foundations in safety, and are associated with contentment and happiness. They are not associated with maladaptive and addictive behaviors. Modern Western culture is obsessed with conflict, competition, and winning—this is hugely problematic for human safety, health, and happiness.

Modern societies and cultures have tended to define masculinity through traits of physical and economic dominance. The egalitarian male disappeared with modern civilization. Population growth created crowding and competition for the best land, fertile with fresh water and pastures for crops and livestock with access to the sea being preferred. This created conflict and demanded boundaries, initially with tribal lands, and eventually with personal property—boundaries do keep us safe. With domestication and loss of the hunter-forager-gatherer and nomadic life the concepts of acquisition, accumulation, surplus, finance, and wealth, and having and being the best, evolved. More stuff was now thought not to be a burden to be carried, but an asset to be managed and protected for more power, prestige, and profit in a world of competition, conflict, and threat. In this world, the self, me, and mine compete with the other, you, and yours, while leaving us and ours far behind. The human race became the race to get the most stuff prior to dying—and "civilized" men took the lead. In many ways, civilization has destroyed civility.

The aggressive, competitive, and stoic archetype of masculinity is the foundation for repression, conflict, rigidity, and abuse—"toxic masculinity"—that has been rejected, properly so, by much of society today. Unfortunately, in the wake of this ongoing cultural shift, boys and men feel apprehensive, rejected, lost, and threatened. The old version of masculinity is fading, and a new version has not yet emerged to replace the old. This loss has created a reactive whiplash with a surge of male supremacy movements and organizations, and hateful antisocial behavior.

Modern culture is not only full of conflict and competition, but full of games and sports that build on these properties and the need for winning. Winning feels good, but remains an outcome born from threat—an unsustainable inflammatory and catabolic state. Games and sports have some value in developing skills and fitness, strategies for resilience and coping with threats, losses, and pain, as well as providing opportunities for cooperation, collaboration, and adaptation. But to have unrelenting conflict, competition, games, sports, and the need for winning dominate the times has become destructive. Culture sustains chronic threat and culture can become dangerous for human wellbeing.

"Achieving gender equality requires the engagement of women and men, girls and boys. It is everyone's responsibility."

Ban Ki-Moon

Boys and men are falling behind girls and women intellectually and academically. There is concern that for every four women who achieve a college degree only three men do so. Falling intellect and academic performance reflects TL, but there is likely more going on here.

Professionally and politically men still dominate, but this too is shifting. For some men professional goals feel unreachable and old social behaviors feel unacceptable. Without a clear path and identity for boys and men, without an alternative construct of masculinity, cults, gangs, militias, terrorist groups, and political parties offering an antiquated path and identity—

a concrete, yet maladaptive, definition of masculinity—are filling in the void. Additionally, within threat, the search for a savior seems more than reasonable. Within chronic threat, the emergence of cults of personality are predictable. In fact, narcissistic traits that have value when faced with an acute physical threat emerge and become profoundly toxic and sustained within the context of chronic social threat.

Unfortunately, those who never find their place in society are subject to the deaths of despair—addiction, overdose, and suicide.[415] Men account for 75% of suicides and 90% of homicides. This is a dangerous place to be.

"As for testosterone, it's gotten a bum rap. Yes, it has to do with aggression, but it doesn't cause aggression as much as it sensitizes to the environmental triggers of aggression."

Robert Sapolsky

The construct that men are naturally more aggressive, competitive, and less empathetic due to higher levels of testosterone is false. Testosterone is contextual. In threat, men with higher testosterone can be angrier, aggressive, competitive, and antisocial. However, in safety, men with higher testosterone can be more compassionate, connected, loving, and social. (Note: women with higher oxytocin levels show similar traits in threat versus safety.)

Men with low testosterone die younger, mainly of heart disease. But is this a causation or a correlation? TVST argues this is a correlation of chronic threat as threat reduces sex hormone production in favor of threat hormone production. Treating a threat phenotype with testosterone could make things worse as testosterone signaling is contextual to threat versus safety and in threat testosterone can promote inflammation. This is a common practice in medicine to treat the symptom and not the cause and it is not infrequent that this process can make things worse. It is important to not just look at the what but also the why—find the root, treat the root—chronic threat.

From an evolutionary standpoint, men are meant for and meant to labor. Labor creates fitness, strength, intellect, creativity, productivity, and satisfaction, and thus meaning, purpose, and value. The devaluing of the trades and labor has contributed to the devaluing of men. It is fair to say academic achievement and wealth are overvalued at the expense of the trades, labor, and men. To propagate the genetic code and the species both men and women were meant and made to labor and deliver, just in different ways.

> *"When wealth is passed off as merit, bad luck is seen as bad character. But poverty is neither a crime nor a character flaw. Stigmatize those who let people die, not those who struggle to live."*
>
> *Sarah Kendzior*

Whether within the halls of academia or within the fields of farming human potential, performance, health, and happiness are tied to relative threat versus safety in the world, but other than the cultural constructs of value and associated compensation, it may be fair to say the trades and labor have more to offer with regards to the total needs of a human being. Yet the lack of respect weighs heavy.

Artificial intelligence is likely to replace white-collar jobs faster than blue-collar jobs. Certainly, AI and robotics will replace labor as well, but humans' evolutionary coding to work and create will not be replaced. Humans are hard wired to do these things, particularly when feeling safe. Celebrate a return to the trades and labor with meaningful wages. Celebrate men and define a nontoxic masculinity based on the known traits of safety. More importantly, cultivate safety. Chronic threat creates "bad" men and "bad" culture. Sustained safety creates "good" men and "good" men serve, produce, protect, and provide safety.

A new cultural identity and definition of masculinity needs to be brought forth. The construct will be important, but a culture of safety will be most important. Let men be men…but within safety, not threat.

LGBTQ Health

> *"…it takes no compromise to give people their rights.…*
> *it takes no money to respect the individual.…it takes*
> *no political deal to give people freedom.…it takes*
> *no survey to remove repression…"*
>
> *Harvey Milk*

TVST also explains the health inequities and disparities seen within the LGBTQ community. In fact, any person or population perceived as different, ostracized, and threatened by the greater community is at risk for poor health. The LGBTQ community is no exception.[416]

The LGBTQ community is representative of the diversity of humanity. The LGBTQ population is present in all societies and cultures across the globe. Yet, in almost all societies and cultures, LGBTQ individuals tend to suffer more abuse, marginalization, prejudice, disenfranchisement, discrimination, inequity, and injustice. LGBTQ individuals' TLs can be higher than the general population, and poorer health follows. Additionally, transgender individuals are at higher risk of suicide attempts and mortality than non- transgender individuals.[417]

Although genotypes are typically (but not always) male or female, life has infinite phenotypes that include straight, lesbian, gay, bisexual, trans, and queer, and everything in between. At a cellular level race, gender, and sexual preference are insignificant. At a cellular level, humans are the same. At a subcortical level, the different is suspicious, be on alert...and unfortunately be biased. But it is at the cortical level that humans create the illusion of significance to race, gender, and sexual orientation through faulty contracts, ideations, constructs, narratives, and beliefs. Perhaps, if biological models are used then it will be easier to move beyond race, gender, and sexual orientation to see the value of life itself.

Policies, procedures, laws, and culture should be built around the concept of safety. If it makes someone feel safe (healthy and happy, too) and does not threaten another then it is for the good. "Victimless crimes" are not crimes at all. These "crimes" are false contracts, ideations, constructs, narratives, and beliefs used to marginalize and disenfranchise selected people or groups of people. This is counterproductive, as it increases the TL in those individuals, societies, and the world—it hurts all of us.

The Declaration of Independence states:

> *We hold these truths to be self-evident, that all [people] are created equal, that they are endowed by their Creator, with certain inalienable rights, that among these are Life, Liberty and the pursuit of Happiness.*

It is frightening to see legislative bodies and courts using cultural constructs, not scientific or constitutional constructs, within their determinations. A good measure of the progress humans are making in moving the world toward safety will be the health of the individuals and populations long marginalized and disenfranchised due to maladaptive contracts, ideations, constructs, narratives, and beliefs.

Race and Racism in Health and Healthcare

"There is no such thing as race –
scientifically, anthropologically. None.
There is just the human race."

Toni Morrison

Consider the fact that the human genome is remarkably similar—99.9+% the same from person to person.[418] Humans appear, act, and are very similar whether in threat or safety. All humans are pink internally, and all humans bleed red. In addition, humans are fundamentally and inextricably connected in action and reaction, whether in threat or safety. Both threat and safety compound within the individual, society, and the globe. All humans have the same wants and needs, goals and aspirations, yet sometimes unaware and unspoken, to be safe, seen, and secure in the world. Nevertheless, safety is declining across the globe. This trend needs to be quickly reversed. Time is running short before this human race is lost.

Threat versus safety plays out in many ways within health care when it comes to race and racial inequities and disparities. Racism in medicine is complex and complicated.

Racism comes in an explicit form. This form of racism arises within cortical awareness and contains social contracts, created ideations, cognitive constructs, symbolic narratives, and held beliefs. Therefore, racism can be addressed to some degree through education, but it cannot be changed with education alone.

Racism also comes in an implicit form. This arises out of awareness of differences within the subcortical predictive codes where biases and prejudice live out of cortical consciousness. With a high TL, humans become suspicious, sometimes paranoid. They are biased to see threats, misinterpret signals to suspect ill will when none is present, and can inaccurately intuit differences and deficiencies in the genome, brain, or character of another. These processes all lead to threat-related social disconnections, paranoid ideations, cognitive distortions, false narratives, and inaccurate beliefs.

"It may be doubted whether any character can be named which is distinctive of a race and is constant."

Charles Darwin

Prejudice—the ability to quickly judge prior to thought—is a natural state when under threat. Prejudice lives more in predictive codes than in cognitive awareness. It is more implicit than explicit. Prejudice was an asset in a world of physical threat. Advanced, but slow, executive functions had little benefit in a tiger attack.

Prejudice is problematic in the modern world where non-physical threats predominate. Humans must be aware of this threat-related physiological and psychological phenomenon. How this phenomenon factors into institutional and clinician biases and misunderstandings and leads to marginalization and neglect within healthcare today cannot be overlooked.[419]

In addition, biases, prejudices, and narratives play roles in who enters the medical profession. Admission processes do not select who is most needed and capable of providing medical care to communities. Rather, these processes and people select those most ready for a medical curriculum at that point in time—a relativity that may not determine who will become the best clinician. This is based on standards biased towards the most traditionally and formally educated—the wealthy and privileged. To be certain, there are barriers all along the way for everyone to overcome on the path of becoming a medical professional. It is hard work. But nevertheless, there are many more barriers for the less privileged and less well-educated. These barriers deter, dissuade, and defeat even the most capable before ever getting started. Competition on an unlevel playing field is the present platform for access to medical training.

Self-bias is an issue as well. Long disenfranchised minorities simply can't see themselves in the role of a clinician, both professionally and socially. Having a comfortable peer group and being safe, seen, and secure within a difficult profession can be the difference between thriving versus surviving, or even succumbing.

Again, competition allows for surviving threats, but cultivation promotes safety and thriving. Safety physiology allows for mental flexibility, curiosity and creativity, and safety defeats threat biases and promotes safety biases. All interested in medicine should be cultivated and mentored. Need, desire, capability, and potential should displace biased academic data. Medicine needs minorities and diversity. This is not as a social agenda, but for the quality of medicine and the health and happiness of the global population.

"The purpose of human life is to serve, and to show compassion and the will to help others."

Albert Schweitzer

Institutions of healthcare are distressful for both patients and providers. Patients fear illness, disease, disability, pain, suffering, complications, and death. Patients may fear practitioners and the very institutions that provide healthcare.

Practitioners fear conflicts, difficult patients, mistakes, adverse outcomes, and lawsuits. In addition, practitioner's TLs are elevated due to the constant monitoring and scrutinizing of their patient satisfaction, production, billing, compliance, behavior, and quality of record keeping. With more threat and burnout, more problematic issues around prejudice arise. Threat-induced prejudice, discrimination, and lack of empathy, compassion, and caring will always rise within the health care sphere given its current culture.

Empathy—the ability to sense what another is experiencing or feeling—is a natural biological state when in safety. Empathy reduces pain and suffering.[420] Empathy is physiologically suppressed in threat states. Connecting, bonding, and empathizing have little benefit in a tiger attack. Notably, empathy is a physiological state, an emotion, that can only be experienced and not taught. Empathy can be defined at the cortex within social contracts, cognitive constructs, and symbolic narratives, but it is not born there. Empathy is an emotion that arises from safety at the depths of the genome, expressed epigenome, and physiology. Empathy becomes a feeling when it reaches the cortex of the brain. Empathy is felt—not learned.

As mentioned above, the more threat and burnout clinicians experience, the less empathy they will have. This reflects biology, not genetic, brain or character deficiency or defect, or malintent. Empathy has been rated as the number-one quality patients are looking for in a clinician. Yet, empathy falls during the years of medical training and beyond in relation to TL.[421,422]

When a patient is afraid, connection, empathy, compassion, and cues of safety—care—may be the most important things a provider can give to that patient. Yet, in states of threat, these qualities are unavailable to clinicians and patients. For a clinician, the lack of empathy, compassion, and caring is the equivalent of practicing surgery with a dull scalpel. Safety is required for empathy, compassion, resilience, caring, successful relationships, and good outcomes.

Racism in medicine also involves a lack of education and inquiry regarding the different cultural and medical issues of minority populations. Missing data plays a role in inequalities and disparities in health care. This applies to women's issues as well. This reflects explicit and implicit biases and priorities. More concretely, however, it is seen in the lack of research and the lack of clinician engagement and curiosity in taking a history, performing exams, providing after-visit instructions, and ensuring follow-ups for minority populations.

Reduced access to care due to lack of resources or lack of trust in the medical system also leads to disparities in health care, health care outcomes, and the health of minority communities. Much of these pieces of racial inequity and disparity are missed within the conversations around racism.

"You can't separate peace from freedom because no one can be at peace unless he has his freedom."

Malcom X

Total TL and threat physiology determine dissolution, catabolism, inflammation, degeneration, destruction, illness, disease, disability, pain, suffering, and outcomes. Below the brain, below the concepts of explicit and implicit racism, lies the physiology of threat. That is the root of the problem. This root extends well beyond the institutions and the practitioners. It reaches to all of society and culture. This massive root system feeds the entire system. It should be said that the root absolutely feeds and affects every cell in every being. The literature bears out the correlation that the higher the TL, the worse health and health care outcomes seen.

Society and culture play an enormous role in illness and disease versus wellness and health. Racism, prejudice, subordination, oppression, and abuse play a part in the equation of the TL and determine physiology. Generational traumas, disenfranchisement, discrimination, inequity, injustice, and poverty all have major health impacts.[423] Notably, high-threat physiology is seen in many indigenous, black, and brown populations with extensive histories of trauma and threat.[424] A minority adolescent experiencing discrimination will have activated TS, TC, and an associated TR with inflammation and altered immune and metabolic function. These children are more susceptible to altered sleep and as adults are much more likely to suffer from metabolic syndrome. In addition, black adults who experience racial discrimination suffer from poorer sleep, increased inflammation, and higher rates of metabolic syndrome. All of these threat issues contribute to poor health and poor outcomes associated with hospitalizations and medical interventions. Again, the body keeps the score.[425]

Black adults are more likely than white adults to live in poverty, which is associated with almost 50% greater premature mortality. Premature death rates in black versus white adults drop from 59% to 0% after adjusting for the social determinants of health. Efforts to address the total TL are essential to address health disparities. A concerted effort needs to be made to address the structural, institutional, political, procedural, legal, and societal determinants of health and happiness, targeting the TL as the underlying causes of premature death among all people.[426] Again, this is particularly salient to those from historically marginalized populations.

Additionally, poor academic performance is not a factor of race; it is a factor of chronic threat. It must be addressed by removing threat and providing safety. Adversity in childhood causes threat-associated changes in brain structures. False narratives around race, along with academic performance and capability, damage individuals. This damage manifests itself in the form of further bias, prejudice, discrimination, and escalating threat.[427]

Racial discrepancies, disparities, and inequities within healthcare are complex, complicated, and multifactorial issues. However, the root of all these factors is threat and the physiologic consequences of threat, to both the patient and the provider.[428] TVST postulates that the disparity in TL is the primary and major determinant of the disparity in health and health care outcomes seen between races. The TL is the target for fixing this problem.

This chronic and endemic disease of racism that plagues the country cannot be completely fixed through explicit interventions and institutional or individual corrective actions. Addressing the implicit—the codes and biases below cortical awareness—will need to be done as well.

There is only one race, the human race, anything beyond that is a false contract, ideation, construct, narrative and belief, including the rationale for classes. Deconstruction of the concepts of race and deconstruction of the ruling class will restore equity to the human race. Additionally, the solution to racism will require major societal and cultural shifts. These shifts can be best achieved by moving people out of threat and into safety. Adjusting the implicit changes the explicit, more than the other way around. Adjusting the epigenome, phenome, and physiology through safety will adjust the entire system to start to resolve racism.

Covid-19

> *"It must be confessed that though the plague was chiefly among the poor, yet were the poor the most venturous and fearless of it, and went about their employment with a sort of brutal courage."*
>
> *Daniel Defoe – 18th century*

In the midst of the Covid-19 pandemic, it became clear that TVST defined who fared well and who didn't with a Covid-19 infection.[429] Those who suffered from mental health issues, isolation, disenfranchisement, discrimination, injustice, and lived in poverty, as well as the aged, fared worse.[430] Those with a higher TL and associated comorbidities suffered more mitochondrial, cellular, tissue, and organ dysfunction and injury, as well as shock and death. It is true those people also had more comorbidities. But these comorbidities should not be regarded as causative of a worse sequela, but as coexisting and correlative with worse outcomes, and a result of a high total TL. Similarly, those who had a higher TL and more intense and sustained TS and TC were more likely to suffer from a protracted post Covid-19 infection course.[431]

TS and in part TC are responsible for the phenotypic shift causing oxidation, inflammation, edema, fibrosis, coagulation, catabolism, degeneration, fatigue, dissociation, anxiety, depression, and cognitive decline in a Covid-19 infection and post-infection recovery. Indeed, many of the symptoms associated with a Covid-19 infection, as well as post Covid-19 syndrome, are easily explained through TVST, TS, and TC (including "cytokine storms"—T2 phenotype and physiology associated with all shock syndromes).[432] Managing more than the viral load, i.e. managing the TL, as well as the TC (MABs preferred over GCSDs) can save cells, tissue, organs, and lives.

Covid-19 allowed us to see all the slow degenerative effects of chronic threat condensed within a shortened time frame of an acute illness.[433] The programmed systemic oxidative, inflammatory, metabolic, immunologic, endocrinologic, dermatologic, neurologic, respiratory, cardiovascular, gastrointestinal, fascial, muscular, psychologic, cognitive, and other changes associated with an acute severe Covid-19 infection and sequelae are more rapid, dramatic, and obvious than in chronic diseases.[434] But they are essentially the same processes. It is illuminating to see the TR play out rapidly in real time.[435] This awareness needs to be applied to chronic diseases.

The chasing of the virus without acknowledging the full sociopsychobiological context of the pandemic goes a long way towards explaining over a million Covid deaths. Despite the investments, technologies, and knowledge, and accounting for less than 5% of the world's population, the United States accounted for 15-20% of the world's Covid-19 deaths (the U.S. also has over 20% of the world's prison population and consumes 80% of the world's opioids—any correlations?). This speaks to both the status of the U.S. medical system, as well as to the cumulative TL of the country. This is alarming—and the alarm has been unheeded.

It should be noted that these numbers do not include the Covid pandemic's precipitation of other noninfectious physical, emotional, social, mental, financial, and spiritual illness, disease, pain, suffering, and deaths. In other words, this number does not account for the collateral damage of the pandemic, including the effects of long-term isolation and financial injuries.

Up to 14% of the U.S. population has experienced Long Covid. At any point in time approximately 5% of the population is experiencing symptoms of Long Covid. People with pre-existing high TLs and associated phenotypes and morbidities are at higher risk for Long Covid. Long Covid is no different in symptoms than some people experience after an influenza virus infections, presents suspiciously like myalgic-encephalomyelitis, chronic fatigue, adrenal fatigue, fibromyalgia, postural orthostatic tachycardia,....medically unexplained syndromes. Long Covid even has significant overlap with some symptoms of menopause. In other words, Long Covid is not specific to the Covid virus and looks suspiciously like a T2 TR.

Long Covid findings are consistent with pre-pandemic research on other health conditions, suggesting that people with lower socioeconomic status have poorer health outcomes and higher premature mortality than those with higher socioeconomic position. Notably, having an income over $200,000 is most predictive of all factors in having a protective effect against Long Covid. These findings should help inform health policy in identifying the most vulnerable subgroups of populations so that more focused efforts are given, and proportional allocation of resources are implemented, to facilitate the reduction of health inequalities.[436]

To prepare for the next pandemic, it will be important to preemptively address threat and safety within individuals, societies, cultures, and the globe. In future pandemics, more than chasing the virus must be done; the longstanding chronic and concurrent acute sociopsychobiological threats must all be addressed. This is particularly vital during acute infections and in the immediate post-infection period for surviving but is also necessary for long term healing, recovery, and thriving.

Summary

"The germ is nothing,
the terrain is everything."

Louis Pasteur

TVST purports to explain most of the illness and disease in the world today. Chronic threat physiology seems to easily explain the spectrum of chronic diseases from psychologic neurologic, endocrinologic, gastrointestinal, genitourinary, vascular, cardiac, respiratory, dermatologic rheumatologic, to fascio-musculo-skeletal diseases, and even cancers.

The greatest concerns, expenses, morbidities, mortalities, and quality of life issues in health care today include:

- Cardiovascular diseases
- Cancer

- Metabolic syndrome
- Obesity
- Type 2 diabetes
- Neurodegenerative disorders
- Chronic pain disorders
- Mental illnesses
- Behavioral disorders
- Addictions
- Violence
- COVID-19 infections (and their sequelae)

Dig deep, and one finds that chronic threat is not at the root. Chronic threat *is* the root of all these afflictions.

To understand what ails humans and the world today, start at the beginning of life on earth with simple single-cell life forms—prokaryotes. Then follow the evolution of life to more complex cells—eukaryotes (nucleated, multi-organelled archaea with endosymbiotic bacterial and viral descendants and remnants). Ultimately, go all the way to the most advanced colony—the most complex, multi-networked, multi-organed, differentiated-cellular, multi-cellular, multi-organelled, eukaryotic apex life form on earth today—humans. The understanding of simple life forms within their connections to and influences from their ever-changing environment—specifically threat versus safety—can help to elucidate the exposomic, genomic, epigenomic, phenomic, physiologic, proteomic, metabolomic, connectomic, biomic, physical, emotional, behavioral, social, mental, cultural, and spiritual changes seen in human beings that differentiate disease and illness from health and wellness.

In fact, it is time to differentiate phenos ("showing" the genetic coding) from pathos ("abnormalities" due to toxins, mutagens, allergens, antigens, pathogens, predators, and trauma). It is crucial to understand the roles that both phenos and pathos play in illness and disease. They can overlap, but most importantly is the understanding where they are different. TVST hypothesizes that most of the chronic disease, disability, pain, and suffering is phenos and not pathos. The question is not what but why, and the answer is threat versus safety.

Most illnesses and diseases can all be clustered into the auto-immuno-inflammo-metabolic category in the sense that their origins are found in the transcription and translation of the individuals' genetic code in response to threat to create a systemic phenotypic shift and a cascade of physiologic change into a defensive state with metabolic programs to support this state. Consider a category of auto-phenotypic illness and disease, but Safety Deficit Disorder could do.

The current models for health care are heavily biased towards pathos, which leads to false paradigms and mistargeted treatment algorithms. Trying to treat the complexity of systemic phenotypic and physiologic variations of a TR as if the problem is a singularity, such as a toxin, mutagen, allergen, antigen, pathogen, or injury is a mistake. In fact, what is occurring is a complex exposome-meets-genome cascade of events, with a multitude of threat-related changes. This focus on treating singularities will seldom work and may cause more harm, with further illness, disease, disability, pain, suffering, and at times, even death.

> *"When we look at the pathophysiology, we look at a disease, it's like looking at a battlefield. We come in and try to understand why that battle was fought by looking at the debris and the bodies on the battlefield. That doesn't tell the story of what was going, what was dysfunctional that led to the battle."*
>
> *Steve Overman*

Having understood the basic concepts surrounding toxins, allergens, antigens, and pathogens for over a century now has not led to finding cures for most chronic illnesses and diseases. More recently the focus has shifted to genetic variables that may account for chronic illnesses and diseases. The thirteen-year long and $3 billion Human Genome Project, finally, elucidated the human genome. The project was hailed as one of the greatest achievements in modern science and medicine. Yet, the elucidation failed to solve nearly all the causes of chronic illnesses and diseases and did not alleviate the pain and suffering of the world. Mapping the genome in many ways appears to have drawn a map of the globe, but without the roadways of connection. These connections are left to be made.

The questions being asked are the wrong ones. The paradigm being used is the wrong one. And the algorithm being followed is inadequate to achieve wellness, health, and happiness.[437] TVST hopes to be the start of a map for the lost.

Many have suggested it is time to disregard the genome in favor of studying the exposome—the environmental influences—within illness and disease. Given that life arises at the interface of the genome and the exposome, it seems wisest to disregard neither the genome nor the exposome. The more prudent course of action is to study this interface, especially within the context of threat versus safety.

This is where the genome is being played by the exposome to create and influence life, much like a pianist plays a piano to create and influence music. What is considered dark or dead DNA that codes for nothing will likely prove to simply be unplayed keys and strings within a given concerto waiting to come alive when the pianist, the environment, demands. The genome without the exposome, like the piano without the pianist, is dormant and nearly inorganic in its absence of life's activities. To study the genome outside of the context of the exposome is inadequate and meaningless. The pianist determines whether noise or music is created.

The human genome has fewer genes than some fruits yet

shows an extremely complex and diverse phenome. Epigenetic coding and the vast exposome allow for this complexity and diversity of the phenome. Perhaps the genome is more of a ukulele within the hands of a talented musician than a piano?

Additionally, there is an exploding plethora of new metabolic and immunologic therapeutics. Purine inhibitors and kinase inhibitors (NIBs) block signaling and metabolic functions involved in a TR. Monoclonal antibodies (MABs) utilize immune mechanisms to block ligands and receptors in various components of a TR. What is missing in this equation is the induction of a SR. Blocking TS is inadequate to complete the cycle of healing. In fact, blocking some components of a TR can impair the signaling of when it is time to transition to a SR and healing (see NSAIDs, GCSDs, and IL6 inhibitors*)***,** and make individuals susceptible to chronic pain, infections, and certain cancers. To heal, find health, feel well, and thrive, SS and a SR are required. The question is by what means can SS be induced? Metabolics and immunologics? Or by more comprehensive and holistic individual, societal, cultural, and global therapeutics and interventions? To date, pharmacologic and procedural interventions have been very poor substitutes for an environment of safety.

"Stopping a hurricane by removing
a drop of rain doesn't work."

Ali F. Oeus

Both at a cellular level and at a whole human level there are choices to be made: Conflict or cooperation. Competition or adaptation. Degeneration or regeneration. Cancer cells or stem cells. Infertility or fertility. Infidelity or fidelity. Incivility or civility. War or peace. Pain or happiness. Pathogen or symbiote. Death or life. Yet, these are just correlates. The fundamental choice is threat versus safety, the root causes and differentiators for illness and disease versus wellness and health.

TVST is a dramatically different paradigm that could shift worldviews and impact all institutions, policies, procedures laws, societies, and the globe. This theory has implications across all sectors and institutions, including:

- Education
- Science
- Health care
- Public health
- Social welfare
- Criminal care
- Law and justice
- Economics
- Business management
- Multimedia
- Foreign affairs
- Defense
- Global health
- The environment

It demands reassessment of all institutions, policies, procedures, and laws. Assessment of these areas in terms of how they are weighted toward the elimination of threat and the provision of safety is needed.

TVST forces us to examine which individuals, communications, behaviors, institutions, policies, procedures, and laws increase conflict and threat versus cooperation and safety in the world. The evaluation of the long-term consequences from the areas of threats and toxicities is also needed.

Can a system stuck in threat physiology—inflamed, catabolic, degenerative, infertile, etc.—be called an illness and disease state? Or is it simply a system doing exactly what it is supposed to do within a state of chronic threat? The system is perceived as maladaptive, but is it? Is chronic threat what is maladaptive, not the system? Is the environment, the exposome, simply too toxic for way too long? When the exposome interfaces with the genome, there are shifts in phenome and physiology. Curing isn't the goal; shifting from threat to safety is the goal.

"What if the theoretical assumptions or the assumptions underlying certain theoretical strategies have been wrong? And in medicine, they have been wrong. So, the notion of symptom clusters, having a pathophysiology and the pathophysiology being genetically caused, those are not deterministic sequences. There's a lot of variation."

Stephen Porges

Clinicians, hospitals, and medical centers do well when they focus on the treatment of pathos and injury. They are not well equipped to deal with treating chronic threat phenos. Nor do they do well with regards to safety and outcomes when overrun by corporate and private equity demands for profits and/or returns on investment.[438] Chronic threat related disorders cannot be adequately treated within the current paradigm. Technically, phenotypic illness and disease cannot be permanently cured, as its potential expression remains within the genetic code—to rear up again as chronic threat returns.

Also, for consideration is the idea that hospitals and medical centers are terrible rehabilitation and healing centers. Much money is wasted, and much harm is done treating phenos as pathos within these facilities. Hospitals and medical centers should be first in treating pathos, but not phenos.

A whole new algorithm is required for treating threat phenos. This may require different facilities and clinicians specifically devoted to the treatment of phenos, with resources aligned for healing, not diagnosing, or providing pills, infusions, and procedures. This concept is a treatment for financial ills as well. The cost of healing facilities is much less than traditional medical facilities. Outpatient safety recovery centers and inpatient safety recovery facilities within a new health care paradigm and algorithm are needed moving forward.

Additionally, criminal care is health care and therefore criminal care should follow the same paradigm as health care, but with more accountability, boundaries, and security. Illness and disease come from the same root as criminality and violence—chronic threat.

These concepts prevent the potential harm done from the misappropriation and misapplication of monies, pills, and procedures.

"I think one's feelings waste themselves in words; they ought all to be distilled into actions which bring results."

Florence Nightingale

It will be important to study both the determinants of threat and the determinants of safety. It would also be interesting to study the different determinants of safety that lend towards health and longevity versus happiness and joy. A balance between individual freedom with a strong economic engine supported by transparent competitive market forces and a robust cooperative and supportive societal welfare system is the homeostatic point to target. From this, a healthy and happy human society and culture should arise. The provision of safety is the mechanism to get there. The ultimate outcome would be health, wellness, and happiness for individuals, society, and the globe.

Everyone should know their Health Quotient—SL in relation to TL.

$$HQ = SL/TL$$

HQ should be another vital sign—or perhaps a vitality sign.

Evolutionary history suggests the greatest leaps forward have not come through threat, conflict, competition, and punishment—the surviving of the fittest. They have instead come through safety, cooperation, adaptation, and restitution—the thriving of the wisest. A case in point is the endosymbiotic relationship between an oxygen-tolerant bacterium and a larger protective archaeon that allowed for a massive explosion of life on Earth, and 2 billion years later birthed the human species. [439,440]

"The threat response only allows us
to survive in the short run.
The safety response allows us
to thrive over the long run."

Ali F. Oeus

Postscript

"Our lives begin to end the day we become silent about the things that matter."

Martin Luther King, Jr.

The Greater Good

As Covid-19 fades into the background, the global public health priorities have recently been identified in an article in the journal *Public Health Challenges* (2023). They include [441]:

1. health systems
2. mental health crises
3. reproductive and sexual health
4. malnutrition and food safety
5. diabetes
6. cancer
7. environmental pollution
8. substance abuse
9. infectious diseases
10. climate change

It is worth dividing these priorities into threats, i.e., poor health systems, malnutrition, food insecurity, environmental pollution, infectious diseases, and climate change, being threats, and then the symptoms of chronic threat, i.e., mental health crises, substance abuse, reproductive and sexual health issues, diabetes, and cancer. TVST argues that if we take care of the threats, then the symptoms of threat will take care of themselves by subsiding or resolving without other interventions as the TL declines. Therefore, precision in identifying causation and correlation becomes vital in the design, efficiency, and effectiveness of any treatment algorithm. Target the root—the threats.

Mental health crises, substance abuse, reproductive and sexual health issues, obesity, diabetes, and cancers are all on the rise, and most notably increasing in younger and younger people. These issues are the red flags of a world plagued with threat, but not the roots of the problem.

I'll suggest that the list of priorities above seems less than complete and a little convoluted. The Top 10 public health issues should only be the threats, not the fallout. The fallout will self- correct with the resolution of threat. With that caveat, consider the top ten threats to be:

1. toxic human contracts, ideations, constructs, narratives, beliefs, and associated culture
2. climate change
3. social injustice/financial inequity
4. lack of safe housing
5. food inadequacy
6. infectious diseases
7. environmental pollution
8. disconnection from the natural world
9. unregulated technologies
10. lack of a safe community

Chronic disease, disability, pain, and suffering grow from

these threats. Attention and resources should be directed at these Top 10 threats.

More specifically, the biggest threats to humans in the world today start with air pollution (include tobacco smoke here), toxins (include alcohol here), motorized vehicles (include humans driving cars here), explosive weapons (include humans operating guns here), media (include humans practicing predatory marketing here), and inequity (include poverty here). These are the relatively low-hanging threats to be mitigated for human health, happiness, and longevity.

Focusing on the public health—population, societal, and global issues—will benefit the health of the individual, not exclude or erode the rights of the individual. Additionally, individuals can still be accountable to their own health. Understand that the paradigm of TVST allows individuals to address the determinants of health within their control, yet TVST requires society to address the rest of the burden that only the collective can change.

The common theme in preventing or treating most chronic disease, disability, pain, and suffering is the removal or threats and the moving towards safety—a simple concept, but complex in its execution. Living without threat is not possible. Threat is part of life, so the goal is to minimize threat and negotiate threat more skillfully. Then we need to seek safety by moving towards the things in life that stimulate safety physiology. Perhaps most importantly, individuals need to do their very best not to increase threat in the world and to provide safety for others, as well as themselves. This is the skill of life.

Individuals have a say in directing their lives, but as genetic expression reflects an organism's environment, other people, society, culture, and nature—all the "others"—play significant roles, too. The focus of healing, health, and wellness needs to incorporate safety-based contracts, ideations, constructs, narratives, and policies for both the individual and society, the greater good. Although we can and should always start in our own little corner, the individual cannot solely determine where they land on the threat versus safety spectrum. Large population care is required for the individual to be safe, healthy, and well. The group needs to be incorporated into the process and the process must be executed through the group. In this case, the group is the global population.

Too much focus can be placed on the individual and the individual's emotions, behaviors, thoughts, and mind to try and affect their health and happiness. I'd like to reiterate that there is no mind-body duality—psychology is physiology. This is an important tenant to TVST, physiology, and all the effectors of physiology, must be kept in mind in defining solutions to chronic disease, disability, pain, and suffering.

The formula for feeling well and being well does not come from our thoughts. Our thoughts flow as automatically as our breath, heartbeat, gastrointestinal motility, and our emotions. At a certain level, even our behaviors are automated. All these things are programmed and driven from the physiology that baths our system on a spectrum from threat to safety. When the environment meets our programming, physiology flows without our cognitive input nor our direct control.

Our uniquely human brain with a massive neocortex takes in our physical and emotional feelings, and our thoughts, to search for protections, connections, and meaning. This cortex produces our created ideations, cognitive constructs, symbolic narratives, and held beliefs which may be helpful—or not, reality—or not, worth holding on to—or not.

Too often we accept our ideations, constructs, narratives, and beliefs as reality and hold to them as sacred even when they are not servicing us well. Many therapeutic approaches attempt to change our ideations, constructs, narratives, and beliefs as a first-line approach to heal an injured soul. They do so before addressing the underlying physiology that drives the biases and content of the ideations, constructs, narratives, and beliefs—the things that drive the formation of a perceived "reality."

This proves to be folly when the challenge creates discordance and conflict, adding to the TL and threat physiology. The total TL is the root of a threat soup that flows through us and baths all our cells. This is the source of most chronic disease, disability, pain, and suffering.

Language and cognitive based talk therapy can only do so much to change our physiology and if not applied to the proper phenotype at the proper time this type of therapy can be a waste of time and resources, if not harmful. We cannot talk or think ourselves out of anger, anxiety, depression, and despair, as these are feelings. They are automated emotions coming from the physiology produced by the interfacing of our genome with the exposome.

In addition, we can learn from the past, but we can't fix the past—learn and close the door. Nor can we control nor predict the future—open the door and keep moving from this moment to the next moment with all possibilities.

Threat phenotypes characterized by reactivity, impulsivity, vigilance, and rigidity, and void of planning, contemplation, judgement, and flexibility are poor targets for therapies that are based in trying to change created ideations, cognitive constructs, symbolic narratives, and held beliefs. Notably, metaphor and paradox can be dangerous within the concrete minds of the threatened.

So where should we start to heal? At the root, at the threat. Threat has many forms, but threat physiology has relatively few forms. Threat comes from pathogens and predators, ecological pollutants and toxins, generational and historical traumas, and predicted, potential, real, perceived, remembered, or imagined injury, lack of resources (food, housing, and financial inadequacy and insecurity, and poverty), social injury (disenfranchisement, prejudice, and injustice), and lack of connection (human, nature, and spiritual).

Chronic threat doesn't just correlate with chronic psychologic, neurologic, metabolic, gastrointestinal, dental, cardiovascular, pulmonologic, dermatologic, rheumatologic, fascio-musculo-skeletologic, and oncologic diseases; it is the cause of most. Addressing the multitude of threats is primary to changing physiology and becoming healthy and feeling well.

The total TL can be diverse and massive. In many cases the individual will not have the energy, finances, influence, or power to lighten their own load. Society and government must help. However, what the individual can attend to should be facilitated. As well, there are many biohacks into our physiology that can help, but fundamentally the TL must be reduced to heal.

A good place to start the healing process is with safe and secure housing, a guaranteed basic income, and access to affordable comprehensive health care including emotional, behavioral, and mental health care, with full addiction services.

Next in the healing process is good, organic, whole, unprocessed, healthy food, with lots of variety, many colors, herbs, spices, and healthy fats. Supplemental vitamins, minerals, and probiotics are reasonable, particularly if good food is a challenge or at the start of in the healing process. The system needs the proper nutrients to fuel the physiology of health and happiness.

In addition, prior to addressing faulty or false ideations, constructs, narratives, and beliefs the system can be engaged to change physiology through signals of safety. Activation of the anti-inflammatory and anabolic pathways through the healing parasympathetic network can be achieved through breathwork and our very senses.

Good smells—such as lavender, rosemary, eucalyptus, and citrus—change our physiology. Visions—sunlight, starlight, moon beams, skyscapes, landscapes—all shift our physiology. Sounds—vibration, resonance, song, and music—have similar healing effects at a deep physiologic level. Tastes—sweet, sour, and spice—reduce inflammatory and catabolic cytokines and activate parasympathetic vagal tone. Touch and stretch—massage and myofascial therapy—activate healing pathways. Heat and cold—saunas and plunges—promote healing and resilience.

Progressive exercise is also fundamental to healing physiology. We were made to move, and exercise drops our resting inflammatory markers, raises our anti-inflammatory markers, drops our resting sympathetic tone, raises our parasympathetic tone, makes us anabolic in our bodies and brains, thus improving fitness, mood, and cognition.

Movement is primary to the brain's physical, emotional, mental, and cognitive health. Physical movement and flexibility facilitate emotional, social, and mental movement and flexibility. Breath work, yoga, tai chi, walking, running, strength training, play, and labor—(yes, labor; it is as important to our physiology as sex)—are all excellent strategies for physical, emotional, social, mental, and cognitive health.

Higher level biohacks, including high intensity training (HIT) and restrictive blood flow (RBF) training, can be applied when appropriate.

Free uninterpreted emotional expression—movement, rhythm and dance; chanting, singing and music; drawing, painting, and sculpting; play and labor—is important to healing, health, and wellness. Labor shouldn't be marginalized or vilified in this equation.

Rest and recovery are essential. Both good and adequate sleep as well as a day off, a sabbath, are essential to physical, emotional, behavioral, social, mental, and spiritual wellness.

Secondary healing biohacks, including medications, massage, acupuncture, vagus nerve stimulation, neuroresonance training, and transcranial magnetic brain stimulation, are perfectly acceptable options when needed.

A good coach can assist to implement these physiologic strategies by helping with processes that may be offline in the threatened, such as organization and planning, with an accepting and empathetic manner of holding to account.

Where does traditional counselling come in? Phenotype and physiology must shift to engage in changing contracts, ideations, constructs, narratives, and beliefs. Talk therapy is tertiary—the cherry on top, not the first step.

The world is in a pandemic of despair and, unfortunately, our medical, emotional, behavioral, social and mental health structures are not set up to adjust human physiology from threat into safety to make us healthy and happy. We have a perception that mental health care and counselling can fix this problem. They can't.

More ACT, CAT, CBT, CFT, DNMS, DBT, FAP, IFS, MBCT, MBT, MI, PCT, PE, REBT, SDT, TFP...just cannot adequately address the physiology of chronic threat, and implementation at the wrong time results in suboptimal outcomes and marginalization of the therapeutic benefits of any modality. When these interventions are properly fit into a treatment algorithm, they can enhance health and happiness, and prevent the missteps that promote maladaptive contracts, ideations, constructs, narratives, and beliefs, but they are not THE answers and proper timing in their use is crucial.

If we need more acronyms, I suggest considering Nondestructive Emotional Expression Therapy (NEET) and Deconstructive Neocortical (contract ideation construct narrative belief) Therapy (DNT) as some things for which we should all seek, practice, and be skilled at.

Emotional, behavioral, social, and mental health professionals have a key role to deliver this type of education and counselling. But who will take care of the nutritional, financial, societal, and spiritual deficiencies and associated despair that may drive up to 80% of our threat physiology?

When will we realize that it isn't a mental health crisis? It is an ecological, generational, historical, physical emotional, behavioral, social, cultural, societal, spiritual and, yes, mental health crisis. But more accurately, it is a crisis of threat.

The solution is safety, not education and counselling alone.

None of Us Are Truly Safe Until We All Are Safe.

I am an old straight educated white male. I am safer than most. I am more privileged than most. Yet, given my background and station, I am still acutely aware that the modern world is a dangerous place, and the most dangerous species in the modern world is the human species, specifically humans in threat. Threat drives rage, reactivity, impulsivity, rigidity, despair, hopelessness, helplessness, selfishness, and asocial and antisocial behaviors all along the threat spectrum axis.

In this modern world, I too remain threatened. It is a world where contaminated needles litter our parks. Pastel-colored fentanyl is found on our playgrounds. The salve of drugs addicts the desperate. Pedestrians are robbed on our sidewalks. Homes are burglarized with occupants in place. People are shot in their cars, supermarkets, churches, and schools. Nationalists target our power grid. Insurrectionists disrupt civil governance. Terrorists flaunt military-style assault weapons. Despots threaten nuclear holocaust. Profiteers destroy our planet and our very health. And the most resourced suffer from the affliction of morbid wealth—a TR from the excess accumulation and storage of money and stuff.

With this global TL, although relatively safe, seen, and secure, I, nevertheless, am threatened.

All this dysfunction and disease has at its root chronic threat. This compounds and produces more threat. Privilege does not spare me. It simply places me at less risk than the average—and at substantially less risk than the marginalized, disenfranchised, discriminated, poor, and homeless.

Selfishly, to be safe, I need the world to be cured of a pandemic of inequity and threat. Inequity is a virus that codes for this systemic pandemic of despair. Similarly, to Covid 19, this societal virus can infect and consume us all, and it most certainly will without treatment, and the more threatened will suffer the most. Inequity and inequality kill.[442] The most vulnerable, most threatened, will perish first. However, I and those like me are not immune.

So, yes, selfishly I am invested in the cure, but the emotional and social part of my human brain wants the cure for others, too—my friends, my family, my children, my grandchildren, the kids I coached, the folks I taught, their children—and by extension, the world.

How can this happen?

Self-help and self-care are not the sole solutions to wellness and health. In fact, self-focus and selfishness are part of a TR. Self-awareness can be misconstrued and lead to the focus on the false-self, when in fact we need physiologic awareness. This is a focus on the real-self in any moment in time and space. However, focusing on self-protection may be the first step in a healing paradigm to deter threat and ensure survival. Yet, we must move beyond this stage to truly thrive. We must reconnect with the world and all the other it contains rather than obsess in the self to thrive. The other must reconnect with us, as well.

We can't control Mother Nature, but we can inflict, disable, and kill her and thus ourselves in the process. Connection to and protection of the natural world is part of the healing of individuals, societies, and the planet. In addition, we must connect with and protect each other in healthy and safe ways to thrive. The structure of society and how well we care for each other plays a large role in where we as individuals land on the threat versus safety spectrum. Safety promotes prosocial behaviors. Threat promotes asocial if not antisocial behaviors. Therefore, it is reasonable to reassert no one is truly safe until everyone is truly safe. It is an impossible utopian goal, but a valuable construct to keep in mind when assessing the world's problems and solutions.

Achieving individual, as well as societal and global, wellness and health requires a level of societal support, connection, protection, and care for all individuals. To do this the society, the group, must be healthy and well. This, in turn, is dependent on the individuals being healthy and well. We cannot escape this bidirectional connection—it is our biology. We must attend to both the individual and the group—it is not an either/or but a both.

"The survival of the fittest" is a dominant and all-too-valued cultural norm that plays to notions of conflict and competition as the sources of success. This toxic construct fails to optimize our health and wellbeing. The surviving fittest can still be tattered and torn.

Humans with their massive neocortex should be able to grasp and gravitate to "the thriving of the wisest," with its emphasis on cooperation and adaptation as a better strategy for living.

With that in mind we can move towards the application of TVST.

"Think lightly of yourself and
deeply of the world."

Miyamoto Musashi

The Algorithm

The algorithm I use for healing reflects hierarchical biological evolution—phylogeny—that reflects both the individual's needs, as well as the group's. Some interventions the individual can execute on their own. Others require societal interventions for supporting the individual as a component of the whole. The whole cannot be healthy if its components are diseased.

The Code

As life arises from the genetic code when assessing a TL, it makes sense to start at the level of genetic expression.

In review, the genome responds to changes within the environment. The genome takes input and stores information that can be reproduced. The genome's output through transcriptive processes is peptides and proteins—these peptides and proteins drive biologic pathways. There is no direct gene to function connection, this is done through the peptides and proteins produced. The genome changes its structure and code in response to changes in the environment and incoming information thus having flexibility in its outformation. The configuration of the genome determines the interfaces available for transcription, while the sequencing of the reading of genes, and the clusters of genes being read determine a vast variety of outputs—the phenome. The genome is wildly pleiotropic and can transcribe many more peptides and proteins than there are genes through a flexible code, changing structure, order of reading, and clusters being read. And this is before even mentioning the many mechanisms of epigenetic alterations and modifications that can change the output of genetic transcription. No solitary gene operates in isolation to do any one thing. Notably, genomic expression is contextual to threat versus safety.

The genome has some deterministic qualities to it, but the genome is not strictly deterministic. In addition, the genome has no direct and limited indirect influences over the naturally occurring properties of nature that are integral to biologic processes and life itself. Given the infinite environmental inputs to the genome, the pleiotropic nature of the genome, the flexibility of the genome, and the lack of specificity of a single gene's function, pursuing genes, mutations, or specific mutagens to cure diseases may be folly.

We are genetically biased towards threat awareness. Epigenetic bias can stack on top of this genetic bias. Generational trauma and threat biases us toward threat coding of the genome and epigenome that is then expressed in threat phenotypes and physiology. This all predisposes us to illness, disease, pain, and suffering. Technologically rewriting the genetic code is difficult. Mechanistically changing the expression of the code is also challenging, but we are getting more tools for this every day. Nevertheless, it is better to alter the expression of the genetic code holistically, naturally, and systemically by building a phenotype, society, and culture of safety. We must decode threat and recode safety. Preventing threat coding and facilitating safety coding is easier and preferable. This must be done today for not only our benefit, but for the benefit of future generations.

Interstitial and Cellular Safety

"I see the whole of humankind becoming
a single integrated organism....
I look upon each of us as I would an individual cell in
an organism, each of us playing their respective role."

Jonas Salk

In review, our cells are, in fact, much like such simple organisms such as bacteria and archaea. All cells in the body participate in processes of defense and healing. The immune cells are mobile and circulating cells, very similar to their eukaryotic (archaea with endosymbiotic bacterial and viral descendants and remnants) ancestors that play a role in these processes of defense and healing. But they do not have exclusivity in TS,

TC production, and threat phenotypes. They are not just defenders; the immune cells are also healers when they are in safety phenotypes.

The cytokines communicate with our cells within the interstitial fluid—the physiologic soup—to regulate metabolic, immunologic, autonomic, emotional, behavioral, social, mental, and spiritual states. TC affect phenotypes across the threat spectrum from fortress, fight, flight, freeze, falter, to faint. The question is how do they do this when these phenotypes vary so much from attacking and assaulting in a hypertensive, hyperimmune, and hyperkinetic fit of rage to curling up on the ground in a hypotensive, hypoimmune, and hypokinetic state of collapse? TVST speculates that TC formations and concentrations, in association with TC receptor affinities, have a large part to play in this multi-phasic, multi-phenotypic, pleiotropy. Receptor solubility and feedback networks likely also play a role in these transitions in phenotype, too. It is at this level that much of the threat and safety responses are coordinated to either defend or heal.

Next within the paradigm of healing comes the physical needs of the individual cell and the collective of cells—the organism—the things that keep the cells safe, well, and healthy. This is where the cells meet the environment (interosome + exterosome = exposome). Internally, it is the interstitial soup that bathes the cells. Externally the epithelial cells and sensory receptors interface with the world.

Pollutants, toxins, and many unique man-made chemicals are threats to us, and all activate a TR. Air pollution alone decreases average life span globally by 2.3 years. Unclean air is comparable to smoking cigarettes and multiple fold more hazardous than unclean water, alcohol, or motor vehicles. Plastics and PFAS, forever chemicals, directly affect the physiology of humans….and the planet. Recently much focus has been placed on PFAS and their effects on the endocrine network, but PFAS are a systemic threat with effects on the heart, liver, kidneys...and the initiators of cancer. For us to activate safety physiology, we need clean air, pure water, and healthy organic whole food within a secure food supply. This requires global ecological, societal, and governmental support to achieve.

In addition, we all need safe and secure housing with a warm shower, warm bed, warm food, and hopefully, if needed, a warm shoulder to lean on or cry on. These things are basic and fundamental physical needs that allow us to thrive. If we want to prevent illness, disease, disability, pain, suffering, and antisocial behavior to facilitate wellness, health, happiness, and prosocial behavior these things should be a given for all human beings on the planet. Other things that keep our cells healthy and happy include rest, relaxation, and sleep, but also activity, play, labor, and exercise, all in balance.

Real food must become a priority for all—humans and cells. Real food, including vegetables, fruits, oils, and spices, lower TS and TC. Good food should be our first treatment towards wellness, health, and happiness. Food is medicine.

We must meet the basic ecological and physical needs of all to fulfill the prescription for the greater good. This greater good extends from every cell within the body to every being on the planet. This serves all individuals, and thus, makes us all safe.

Autonomic Safety

> *"You might not want to feel needy...but your nervous system is extremely evolved to drive you to seek safe emotional connection with others...and has millions of years of a head start on your wish to be above that."*
>
> *Allyson Dineen*

The autonomic level is most associated with the slender brainstem that sits below the bulbous and undulating upper part of the brain and just above the spinal cord. The brainstem is phylogenetically older than the cortical brain above and 100s of millions of years older than the human neocortex. It is within the brainstem that many automatic, nonvolitional, functions of the brain and body are quickly coordinated to survive, thrive, and propagate the genetic code. Under severe threat, neural dissolution can fall as low as the level of the brainstem and spinal cord, where reflexic behaviors dominate, and social and mental functions are relatively quiescent to absent. The autonomic neural functions bidirectionally modulate cellular functions, but do not completely supersede the autonomous functions of the cells.

The autonomic neurologic level modulates such things as respiratory (diameter, drive, and rate), cardiac (contractility and rate), vascular (diameter and pressure), gastrointestinal (motility and secretion), genitourinary (bladder and reproductive organs), dermatologic (oil, sweat, and hair), endocrinologic (metabolic) and immunologic (defense and healing) functions across the spectrum of phenotypes from threat to safety.

The parasympathetic network and sympathetic network make up the autonomic neurologic level and have their supporting nuclei within primitive brain structures where the internal and external afferent sensory networks via cranial and spinal nerves transmit their data on the status of the exposome in threat versus safety. The afferent component of the autonomic level is much larger than the efferent component indicating the importance of the connection to the exposome and the collecting of information as to the status of the exposome. Efferent signaling can rapidly modulate the system over distance to help coordinate a program for surviving and thriving to protect and propagate the genetic code.

The parasympathetic network largely goes offline when in

threat, while the sympathetic network mobilizes for action, whether in threat or safety. Airways dilate, respiratory drive and rate go up. Heart contractility and rate go up. Blood vessels to the heart, lungs, muscles, and primitive brain dilate, while blood vessels to the neocortex, gut, and reproductive organs constrict. Hair may stand on end, and oil and sweat glands become active. Neurologic sensory inputs, reflexes, and muscle tension increase. Vision narrows and focuses. Cortisol and adrenaline increase, and through catabolic processes, fuels are mobilized. Immune functions are activated with associated increased inflammation, oxidative stress, and trapping and clotting abilities.

In extreme threat the sympathetic network will downregulate, while the parasympathetic network largely remains offline. In this state the autonomic neural level starts to give way to the autonomous cellular level. Airway constriction, lower respiratory drive and rate can follow. Heart contractility and rate fall, and the heart is susceptible to autonomous rhythms—arrhythmias. Blood vessels can become hypotonic and less reactive with low blood pressure and postural blood pressure drops upon standing. Syncope/fainting is common here. Low appetite, poor digestion, gut stasis, bloating, constipation alternating with loose stools becomes the pattern. The skin may become pale and dry. Reproductive functions are not a priority. Metabolism is lowered to preserve tissues. Energy is low in production and presentation. Neurologic sensory inputs are dulled, social drive drops, and cognition becomes foggy, all while autonomous cortical electrical activity can be seen in the form of seizures. Immunologic functions are depressed. Endothelial and epithelial junctions become loose and leaky spilling fluid, protein, and cells into the interstitial spaces, congesting the brain, loosening stools, and pooling in limbs.

The parasympathetic network reverses these phenotypic changes of threat and can be biohacked to give the physiology of safety with lowered inflammation, lowered oxidative stress, increased anabolism, and the restoration of tissue, organ, and systemic functions. But biohacks are not curative—safety is curative.

Stimulation of the parasympathetic network can be done through rest, relaxation, breathwork, yoga poses, pleasant sensory stimulation (smells, tastes, sights, tones, vibration, warmth, cold plunge, touch), gastric and rectal filling, sex, creativity, emotional expression, social connection, spiritual connection, and awe. Less so through mental engagement.

As well, direct vagal nerve stimulation can activate parasympathetic tone. Mechanical, electrical, and ultrasonic vagal nerve stimulation have all had some success in changing phenotypes and physiology, even controlling cytokine storms. More research is being done every day looking at these vagal stimulation devices and their role in managing threat-related diseases. As they will not be curative in that they do not directly address the TL, they will have some limitations. They will have to be used chronically to control symptoms if the TL is never addressed. Another role they may have is to regulate the system enough to allow other more long-term and hopefully curative strategies to be implemented within a holistic and comprehensive treatment algorithm. Vagal stimulation could be a low-no risk jump start to help engage safety-based healing processes.

The major signaler within the parasympathetic network is acetylcholine. Acetyl and choline molecular groups are readily available from mitochondrial oxidative phosphorylation processes associated with safety, but limited when in threat. Acetylcholine agonists may be another avenue to biohack the physiology of safety. Recent research has demonstrated that acetylcholine agonist signaling of safety can control or reverse the symptoms of schizophrenia and are therapeutic in Alzheimer's disease. This would appear to be a much more therapeutic and appropriate pharmacologic intervention than the use of dopamine blockers for controlling emotions, behaviors, and thoughts in these diseases. The latter provides a pharmacologic lobotomy of sorts, as opposed to truly controlling the disease, and dopamine inhibition may prevent healing.

In general, and frequently unknowingly, medicine has focused more on blocking threats and threat signaling, while ignoring the signaling of safety and healing. There is work to be done in this area at the cellular, autonomic, neurologic, endocrinologic, and immunologic levels. For the entrepreneurs and capitalists, there is plenty of opportunity here for power, profit, and prestige. For those less inclined, there is also plenty of opportunity here for the provision of safety within the world.

Repression, Constraint, Suppression, Distraction, Numbing

> *"The challenge of modernity is to live without illusion and without becoming disillusioned."*
>
> *Antonio Gramsci*

Before moving on to the emotional level of the algorithm, let's take a brief revisit of our mechanisms of emotional repression—the prevention of the physiology bathing our cells from reaching our cortical awareness.

Control is paradoxical in that the more control we seek in the world, a world that is largely outside of our control, the more illusions of control we adopt, and the more reckless we become. Chronic illusions of control extinguish safety.

Illusions of control are a product of the distinctly human neocortex that houses complex social contracts, created ideations, cognitive constructs, symbolic narratives, that lead to held beliefs. It is these functions that we frequently are referring to when we refer to our "minds." Arguably, our "minds" have less to do with consciousness and more to do with illusion. In fact, the "mind" itself is an illusion or a false ideation, construct, narrative, and belief. The concepts of mind, the static self, and normalcy are illusions of control. Chronically living within past, present, or future clouds of these illusions prevents knowing life, at all.

Illusions of control present themselves in personality strategies or styles, and sometimes disorders. Personality types are associated with specific gene clusters of expressions—an epigenotype to be expressed in a phenotype and then systemic physiology. These physiologic mechanisms and personality styles or strategies can give us the illusion of control to quell our angst and sense of threat. These illusions of control used chronically increase, not decrease the TL. Examples of illusions of control include:

External illusions of control:

- Victimist – control others through pain, suffering, the need for help, and victim stories.
- Deceptionist – control others through deception, manipulation, and stubbornness, presented under the façades of concern, worry, and niceness.
- Goodist – control their external image through niceness, good acts, sacrifice, and martyrdom.
- Oppositionalist – control by refusing to accommodate another, and through defiant and antisocial behaviors.
- Narcissist – control others through domination, manipulation, and the obtaining of prestige, power, influence, and wealth.

Internal illusions of control:

- Somatoist – control over the body and bodily functions by holding their bowels, limiting eating, purging, exercising, and cosmetic procedures.

- Stoicist – rigid self-control, denial of emotion feelings and physical, social, mental, and spiritual pain to create an illusion of toughness and invulnerability.
- Perfectionist – maintain order through vigilance and obsessively doing everything correctly and compulsively, then checking that everything is right.
- Heroist – seeking achievement and creating scripts of heroism.

External illusion of control strategies tend to align with a T1 phenotype, whereas internal illusions of control strategies tend to align with a T2 phenotype. Similarly, sociopathy tends to align with a T1 phenotype, whereas psychopathy tends to align with a T2 phenotype. (It should be noted that there are gene clusters associated with safety phenotypes and such things as empathy, creativity, optimism, and self-transcendence, as well.)

"Painting is an illusion, a piece of magic,
so what you see is not what you see."

Philip Guston

The social context and social implications of these styles, strategies, if not disorders cannot be understated. A social death can be equally, if not more terrorizing, than the actual idea of annihilation in a physical death (many suicides are proceeded by a sense of a social and spiritual death). These styles and strategies are used not only for restoring a sense of internal control and equilibrium, but to re-enforce the image of the constructed false-self, to maintain an illusion of meaning, purpose, connection, and social safety within cultural norms. The concern arises when these styles and strategies go beyond simple illusion to delusion and frank deception.

We must recognize that manipulation and deception run hand in hand with the biblical form of evil (TVST regards evil simply as an antisocial threat phenotype). Where the belief in the deception, the belief in an unreality, enters is where these styles and strategies become toxic and manifest as maladaptive strategies and frank disorders.

Think of these secondary psychologic diagnoses as the branches of a tree that have arisen from the trunk of the primary psychologic diagnoses (IED, anxiety, depression, BAD, PTSD, schizophrenia.......) of which both trunk and branches have their origins in the root of threat.

As well, people deny, repress, constrain, and suppress emotions, behaviors, and thoughts, thus blocking the information within our physiology, our drives and feelings, from reaching cortical awareness through obsessions, distractions, and addictions such as working, exercising, shopping, gambling, sexing, or just perpetual motion.

Addictions can include excessive working, exercising, shopping, gambling, sexing, **and** substancing, and we can get dependent on or addicted to our illusions of control, as well. At a certain level, with all these activities, dopamine is released. At a minimum, this provides temporary relief from distress, if not frank pleasure, to reenforce the behaviors. Feelings of relief and pleasure can be substituted for the healthier feelings of contentment and happiness in these scenarios.

Additionally, some simply self-medicate to dissociate and numb unpleasant or intolerable emotions and thoughts and their associated feelings. Being unaware of these processes—distraction, soothing, and numbing—can result in other disorders, a slide into addiction, or a fall into a medical, financial, interpersonal, social, or criminal crisis.

Think of the obsessions and distractions as the leaves on a tree. Also think of the addictions as the buds on a tree. These grow from the branches of the secondary psychologic diagnoses that arise from the trunk of the primary psychological diagnoses that all sprout from the vast root of threat. The fruit from this tree of threat is suffering.

Many will use combinations of illusions of control depending on their phenotype in that moment, i.e., multiple phenotypes can bring on multiple isms—victimism, deceptionism, goodism, oppositionalism, narcissism, somatoism, stoicism, perfectionism, heroism, with distractionism and/or addictionism—to avoid the intensity of their aversive impulses, feelings, and thoughts.

Threat, TS, TC, and the associated TR are the root of all psychologic disorders. The "ism," the diagnosis, and associated style or strategy in and of itself is less relevant. The threat is what is important. In dealing with people, or in trying to heal people, we get too caught up in the style, strategy, or disorder and an accurate "check all the boxes" diagnosis (see the DSM) of which none are particularly useful or helpful. We fail to chase the "isms" back to their root—chronic threat.

TVST suggests that stepping away from the DSM and its utilization of descriptors and symptoms without a concept of causation to a system that describes phenotype may be a better evaluative mechanism for understanding neuropsychological disorders and directing appropriate treatments. For example:

Acute Threat States (ATS)

- T1a – Fortress/Protection
- T1b – Fight/Anger
- T1c – Flight/Fear
- T2a – Freeze/Caution
- T2b – Falter/Sadness
- T2c – Faint/Collapse

Chronic Threat States (CTS)

- T1a – Fortress/Cultivation
- T1b – Fight/Agitation and Opposition
- T1c – Flight/Anxiety
- T2a – Freeze/Inertia
- T2b – Falter/Depression
- T2c – Faint/Despair

Chronic Threat Spectrum Disorders (CTSD)

Feature Categories

- a – anxiety
- b – bipolar
- c – conversion
- d – depressive
- e – energized/hypermetabolic
- f – faint/frail
- g – grandiose
- h – hypokinetic
- i – inattentive
- j – jacked/impulsive
- k – hyperkinetic
- l – learning/cognitive
- m – manic
- n – numb
- o – obsessive-compulsive
- p – psychotic
- q – identification
- r – rumination
- s – sleep dysfunction
- t – temper/anger
- u – somatic
- v – visceral
- w – asocial/withdrawn
- x – antisocial/aggressive

- y –
- z –
- ?

Trauma Categories

- Behavioral
- Childhood
- Cultural
- Ecological
- Emotional
- Financial
- Generational
- Historical
- Institutional
- Mental
- Physical
- Social
- Societal
- Spiritual
- Technological
- ! – indicates severe
- # – indicates repetitive
- * – indicates chronic
- ?

This system would thus be inclusive of all primary neuropsychological spectrum disorders including things like ADD, ADHD, Autism, BAD, PTSD, and Schizophrenia, and it would define the phenotype(s) that is most prevalent in the moment for targeted treatments, while never losing track of the causation—threat. This method based on both physiological and psychological presentations to categorize patients seems far better than clustering descriptors and symptoms that are unconnected to a biologic foundation. This system of classification would be more precise and more efficient, plus it would avoid the labeling associated with the constructs of intent and will, or genetic, brain, or character defects or deficiencies. We love to label, condemn, and debate the gene versus brain versus character defect to account for the "check the boxes" diagnosis when what we sense and debate is just not a reflection of the reality of the biology, and it is not relevant to treatment. Dealing and healing starts at the root, not at the more visible trunk, branches, leaves, buds, and fruit, where all the "isms" have grown. The proper diagnosis is "chronic threat." The goal should be to find those threats and treat them properly.

Emotional Safety

"The idea that you have to be protected
from any kind of uncomfortable emotion
is what I absolutely do not subscribe to."

John Cleese

Moving from the autonomic level, we next need to heal at the emotional level. Raw emotions are generated within the primitive brain at the top of the brainstem, the midbrain, where monoamine neurotransmitter reservoirs are housed. This level is where not only mood, but also movement and motivation are generated. Emotions represent our physiology along the spectrum of threat through safety. Once sensed at the level of the cortical brain, they become our feelings. This physiology and neural activity must be felt, integrated, expressed, and safely acted upon for humans to be whole, healthy, and well. Safety allows for this process to be completed. When safe, we can be vulnerable with others and free to fully feel, integrate, and express this physiology, these emotions.

Emotions are meant to move us. Just as we saw gravity

crushing space-time into blackhole singularities, repression crushes emotions into blackholes of numbing despair. Neither allows light or life to escape. Space-time is dependent on movement, and so is life. The loss of movement collapses space-time on a cosmic level and collapses life on an individual level. This need for movement is not just physical, but includes the movement of our emotions. Our physiology is dynamic, and stasis is deadly.

We must have permission and feel safe to express the physiology that is circulating through us. On the contrary, repressing these feelings is toxic. What we resist persists. Repression is sneaky. It lies hidden, yet growing, much like early cancer cells, to finally reveal itself in frank disease states. Leaning into aversive emotions without judgement is important in the treatment of disease, disability, pain, and suffering.[443,444] These uncomfortable emotions are warning us danger is near, so do something—they are our friends and protectors. Chronic judgement, constraint, suppression, and repression are not our friends and are toxic. A fundamental skill humans need is the ability to recognize, acknowledge, and integrate, even play with, our emotions and feelings to achieve health and happiness. This is overlooked and undervalued within our culture.

Finding safe places to express our emotions, whether it be privately, with a safe other, or within a safe group is essential.

There are many forms of expression that help to serve this purpose—rhythm, percussion, vocalization, movement, drawing, and writing to name a few. Not all expression is cognitive, verbal, and language based. All these expressive activities have been demonstrated to be of benefit to physical emotional, behavioral, and mental health.[445,446] Starting with the simplest and purest forms of expression is imperative prior to investing in the expressive arts such as music, song, dance, theater, art, craft, prose, or poetry. Expression must come prior to formal ideation, construction, and narration, but these are natural progressions from expressive therapy. Expression and creativity lead to better physical, emotional, behavioral, and mental health. Safety leads to expression and creativity.

The humanities change our physiology, give us empathy and creativity, and make us happier and healthier human beings.[447,448] However, at the entry level of expression, there should be no goal, meaning, purpose, or end product from expressive exercises. There must only be acknowledgement, integration, and expression of the emotion itself. This is where the healing occurs.

Recognize the difference in expression and narration. Narration can be an illusion of control to repress expression. Expression has a purity that narration does not, and expression can be separate from the "sapiocortex" while narration can not. Expression is more important to human health than narration.

Behavioral Safety

The basic phenotypes of behavior include approach, retreat, and don't move. There is some specificity to threat versus safety in how these basic phenotypes look. In threat, approach is fortress or fight, retreat is flight or hide, and don't move can be freeze or faint. In safety, approach is bond and breed, retreat is feed and digest, and don't move is nest and rest.

Much like our emotions and connected to our emotions, our behaviors frequently arise from the physiologic flow generated by the exposome interfacing with the genome. All of this can be initiated prior to cortical awareness. This physiology creates our raw emotions within the midbrain, most notably at the periaqueductal grey, prior to the emotions being recognized as feelings at the level of the cortex. This same physiology initiates motivation, motion, and movement, i.e. behavior, also starting in the midbrain these impulses travel up the striatum to eventually reach cortical awareness.

This is not to say that the flow isn't bidirectional in that the cortex can choose behavior and choose to constrain or inhibit behavior. But it is to say that the impetus for behavior is present without cortical involvement. When the neocortex is relatively offline, such as in threat states, the exposome meeting the genome disproportionally determines more of our behavior than we do. This is particularly seen when we are hypermobile/hyperkinetic and "out of our minds" in a T1 phenotype, and less so when we are hypomobile/hypokinetic and "stuck in our minds" in a T2 phenotype. Both are considered not unconscious, but nonconscious states.

So, when in threat, in nonconscious states, how much of our behavior are we responsible for or should we be responsible for?

Arguably, very little. In threat states where we are coded, wired, and programmed to be reactive, impulsive, aggressive, explosive, oppositional, vigilant, paranoid, obsessive, rigid, frequently antisocial, and self-focused, self-control is ironically very difficult.

"Accountability is a difficult act of love needed to restore safety...and people."

Ali F. Oeus

Even if we aren't totally responsible for our behaviors this doesn't mean we should not be held accountable to them. In fact, it means we must be accountable to our behaviors, as this is the path to restoring safety. For antisocial behaviors methods of setting boundaries and holding to account must maximize the concepts of removing threat and increasing safety for individuals, societies, countries, and the globe. To do less is at best enabling of bad behavior, and at worst, particular to isolate and punish, only increases the TL and worsens the behavior.

Social Safety

"There is simply no pill that can replace human connection. There is no pharmacy that can fill the need for compassionate interaction with others. There is no panacea. The answer to human suffering is both within us and between us."

Joanne Cacciatore

The next level to consider beyond our emotional network is the social brain that sits above and in front of our emotional brain. All organisms have sociality traits. However, humans have an extremely high evolutionary need for relationality, sociality, cooperation, and collaboration to enhance survival. Humans are relatively weak and slow. Group efforts, including group strategies and group defenses, are essential to the survival of the species.

This biological need for sociality demanded an expansion of the human brain devoted to prosocial functions. This prosocial brain expansion in part accounts for our large heads and prominent foreheads not seen in other species.

Humans' need for sociality drives a fear of a social death, perhaps more feared than a physical death itself, as in the wild, individual humans were incapable of surviving by themselves for very long. Emotions evolved that drive the behaviors of sociality, cooperation, and adaptation.

Defensive states tend to focus on the fear of loss and the protection of the physical self, territory, resources, reproductive mates, and offspring—the false-self too. However, unrecognized to most of us is the fear of, or worse the experience of, a social death that also drives defensive states.

A social death can precede and initiate illness, disease, pain, suffering, and a physical death. Humans are a social species. Life is social.

We have seen that even simple life forms, including ancient bacteria and archaea, are social. Even peptides and nucleotides want to form aggregates, and at times, form intimate condensates with each other. Viruses appear as simple genetic code housed within protein shells that straddle the continuum from inorganic particles and cellular life forms, but they too demonstrate social and antisocial behaviors in the form of cooperation and competition...they even cheat on other viruses. Viruses are not lonely particles, but social pseudo-lifeforms. Primitive simple life forms use social connection to improve their chances of survival and procreation. Humans must, as well.

Chronic social defeat is associated with both depression of the human, but also all the cells in a human—leukocytes, endotheliocytes, epitheliocytes, adipocytes, myocytes, osteocytes, gliocytes, and neurons. Advance distress and elevated TS and TC cause autonomic (parasympathetic and sympathetic) suppression, as well as hypothalamic, pituitary, adrenal (HPA) suppression. Monoamine and cortisol production and receptor activity can be down regulated by threat, TS, TC, and an associated TR. In chronic social defeat, cortisol's function to regulate inflammation is lost. This state is characterized by inflammation, oxidative stress, catabolism, withdrawal, fatigue, depressed metabolism, depressed immunity, and depressed mood.[449]

Lack of social connection is associated with lower brain volumes, cognitive decline, and dementias, as well as all other threat related disorders.[450]

Basic biology demands safe social relationships and institutions, as they are critical to wellness and health. Individuals should gravitate to safe people and safe social structures to limit threats and improve overall wellness and health. Humans have to be aware that this social drive, this need to belong, also puts us at risk for prescribing to toxic cults with foundations of threat. We must be aware of these inviting wolves masquerading as lambs.

A paradox arises in that we are biologically driven to be social to survive, but the more we crave social connection, the higher the fear becomes of a social death. The image building we do and the illusions of control we deploy in the hope of avoiding a social death can paradoxically leave us fearful of sociality, unappealing for social engagement, as well as isolated and lonely. When the social world feels threatening, the dilemma is whether to hide or seek—neither may fulfill the needs to be happy, healthy, and well. This chronic conflict amplifies the TR.

Note: The modern concept and contract of being social is highly biased towards approach behaviors and language-based sociality with the sharing of ideations, constructs, narratives, and beliefs—sometimes incessantly. However, sociality runs much deeper and involves physical, emotional, behavioral, and spiritual connections incorporated into joint activities of exploration, collaboration, creation, and recreation that are void of the need for human language and narrative. Being together in safety is enough to fulfill the human need for sociality. Understanding that not talking is not necessarily asocial as in safety even deeper connections can be found within silence, and excessive language and narrative can be a barrier to these connections. T1 phenotypes may be talkative, while T2 phenotypes may not. S1 phenotypes may be talkative while S2 phenotypes may not. Consider, the apparent extrovert may be in T1 and the introvert in T2…or reciprocally in S1 or S2. Talkativeness is phenotypic and contextual; it is not a good marker of sociality. This concept is important within models for healing.

"Forgive us our trespasses, as we forgive those
who trespass against us."

Lord's Prayer

Society should work to form safe institutions and a culture of safety to enhance both societal and individual wellness and health. This is where boundaries and the ability to set and respect safe boundaries becomes crucial.

People under threat have difficulty with boundaries—this is a threat response, not a genetic, brain, or character defect. They may need boundaries set for them to protect them and others. It is also important to recognize the reciprocity of boundaries. We must establish our own boundaries, and at the same time respect the boundaries of others—the latter being the harder to do. Accountability to the violation of another's boundaries increases safety in the world. Punishment for violations increases threat within the world.

The concept of boundaries—the right to have them and the requirement to respect them—are fundamental to social safety and a safe society. Proper social contracts keep us safe, healthy, and happy.

People in threat phenotypes tend to be antisocial or asocial. Threat is hard on relationships. T1b phenotypes tend to be forceful, aggressive, and disregard boundaries. T2 phenotypes tend to be inert, passive, and have difficulty setting boundaries. Threat phenotypes can frequently find each other, but a healthy relationship is difficult to maintain when both people are in threat and unable to coregulate with each other. In the U.S. 40-50% of first marriages and 60-70% of second marriages end in divorce. Threat and safety inventories may be predictive of the failure or success of a relationship, and it seems advisable not to seek a relationship to be the cure for life's ills. The key is to seek safety so as to be able to be a good partner within a relationship.

"And stand together yet not too near together:
For the pillars of the temple stand apart,
And the oak tree and the cypress grow
not in each other's shadow."

Kahlil Gibran

A big step in being safe with another is to have good boundaries and to respect boundaries. Boundaries make us safe, but they shouldn't make us isolated. Isolation is a behavior of defense, a programmed response to threat—hide to be safe—not to be confused with loneliness, an emotion that drives us to connection. Isolating is a behavior associated with threat, but it is also a behavior of safety. Humans have a need for solitude to rest, recover, and restore themselves, thus to be resilient, as well. In this respect, isolation can be healthy.

Isolation in threat can be adaptive acutely, but it can be maladaptive chronically, and can lead to loneliness. Loneliness is physiology. It is an emotion of distress and despair that drives us to seek connection. Loneliness is a feeling of threat, and its underpinnings are the physiology of threat. Loneliness is a primitive drive, as humans cannot physically survive alone; do something different or perish. Loneliness is a feeling designed to move us from isolation to connection.

Therefore, TVST would predict an association between chronic loneliness and chronic threat-related diseases, disabilities, pain, and suffering. Loneliness is calculated to be as toxic as smoking fifteen cigarettes a day and is associated with anxiety, depression, increased DMN activity, higher alert levels for social cues, increased hyperphosphorylated brain protein accumulation (amyloid, tau, synuclein), decreased brain connectivity, decreased cortical volumes, neurodegenerative disease, dementia, obesity, diabetes, heart disease, increased risk for strokes, worse outcomes from Covid-19...[451] Whereas, connection to the external world counters all of this.

One in three Americans today complains of loneliness—an epidemic. Notably, the Scandinavian countries known for better social and societal support structures and higher health and happiness also suffer from less loneliness.

A dilemma arises when the world is full of threats and human programming promotes hiding. Loneliness follows. Yet long-term isolation in threat is not survivable, either. This dilemma produces conflict at a systemic physiologic level. The question is whether to not move and hide from the other versus to move and seek the other. If both are scary, this is a state of contradictions, conflict, and anxiety.

To fix this conflict, threat must recede, and safety must proceed. There is no other way.

When seeking human connection is fraught with danger in a world filled with agitation, interpersonal conflict, and the weapons of social media—the world is unsafe—we must seek different ways to find connection.

The world has changed from our evolutionary past. Our physiologic drive for a belonging persists. Our physical dependence on the "tribe" has declined, as we are no longer in the wild. Yet, we are still driven to need each other. Sometimes to initiate healing, it helps to find distance from the threats of others and all the modern collective threats through other means of connection. Addition by subtraction can be of benefit at least in the short run—selectively disconnecting from humans and technologies and reconnecting with available sources of wellness, health, and happiness is necessary. Arts, music, animals, and nature are all grounding and connecting.

Additionally, outside of the traditional sensory inputs of smell, sight, hearing, taste, touch, vibration, pain, and temperature, in the absence of any direct communication, language, and narrative, or even any cues of threat or safety, brains resonate and synchronize with each other.[452] How this occurs is not clear. Traditional physics and electromagnetism may contain a clue, but quantum physics may ultimately hold the answer. This should not be a surprise, given the energies of the universe and the physiology and behaviors of ancient simple single-cell organisms billions of years ago, but this connectivity isn't our perception of reality in a threat-based world. Simply to be safe with someone may have tremendous therapeutic benefit, without even saying a word to each other. Allowing distress to subside without trying to fix things will reduce an individual's TR.

Presence can move physiology from threat, disconnection, and loneliness to safety, connection, and resonance even within silence. Perhaps, in this space and time, and the absence of internal and external noise, one can find awe and its healing nature. As threat physiology recedes and safety physiology proceeds, the expanse of human sociality can be restored, or at least resilience attained, to then survive the toxicities of the world today.

In healing, seek connection to the other beyond humans and beyond human contracts, ideations, constructs, narratives, and beliefs, too.

Mental Safety

> *"We are healthy only to the extent*
> *that our ideas are humane."*
>
> *Kurt Vonnegut – Breakfast of Champions*

The emotional brain sets a platform for the social brain. The social brain similarly provides a platform for the mental brain. The mental brain is where ideas, concepts, and constructs are generated and formed. This is the newest part of the human brain that accounts for our massive foreheads and awkward heads, but also our creativity, intellectual acumen, and advanced executive functions.

Note: Some tradeoffs with other brain areas may have been necessitated for this evolutionary step to occur. A major limiting step in perpetual brain growth and evolving functions is the birth canal. A compromise between brain capacity and head size and the survival of the mother and child needed to be reached for the species to flourish. We can only get so smart given this limitation.

The social brain and the mental brain are major components of the neocortex, and they are connected to each other. Both interface with the emotional brain. Our social contracts and mental constructs can have an inhibitory effect on the emotional and behavioral brain—emotional and behavioral repression and constraint. Emotional repression and behavioral constraint are acute survival strategies, but chronic repression and constraint becomes maladaptive and toxic.

Feeling mentally safe is a challenge. In general, we tend to undervalue our emotions and what they are telling us about our reality and overvalue our thoughts and what they are telling us, sometimes about unreality. Based on random thoughts we create so many ideations, constructs, narratives, and beliefs within our heads that may or may not reflect reality and can activate our very real TR. Many of these ideations, constructs, narratives, and beliefs are directly related to our need and desire to belong and connect within society—to be socially safe.

Many of these constructs are learned and frequently involve social contracts, ideations, narratives, beliefs, and norms. These include all our "need tos," "have tos," "shoulds," and "musts," as well as the ideations, constructs, narratives, and beliefs regarding what is required to be "valuable," "meaningful," "purposeful," or "successful" within the culture.

"Can you remember who you were,
before the world told you who you should be?"

Charles Bukowski

These thoughts, ideations, constructs, narratives, and beliefs can become maladaptive over time. Most constructs of the self are false and illusionary. The ability to recognize that these ideas are just constructs, not reality, is important to feeling safe, along with fostering healing, wellness, health, and happiness.

So how does one avoid being trapped by conflicting cultural contracts, ideations, constructs, narratives, and beliefs? How does one still fit into the culture and society that we arguably need to survive and thrive? The answer is the provision of safety to others. The provision of safety to others fosters acceptance and overcomes any concern of atypia within the culture.

Meditation and the practice of not reacting or clinging to thoughts, ideations, constructs, narratives, and beliefs can be helpful. Mindfulness may be inaccessible when in threat, as when we are trying to predict the future from the past, the present moment is elusive. Survival trumps neocortical presence. Interestingly, a mindfulness practice has been associated with less social behaviors and a loss of empathy.[453]

Cognitive behavioral therapy, with deconstructing and reframing, or reconstructing faulty constructs with more adaptive constructs and behaviors, can also be helpful. However, in the severely threatened, these processes may be unavailable. Mindfulness, meditation, and cognitive therapies can help. Yet, they minimally feed the needs of a severely threatened soul and are only a fraction of what is necessary to feeling and being safe within the world.

With little evidence that the traditional strategies of mindfulness, meditation, and cognitive behavioral therapy significantly improve happiness, it is consoling to recognize that gratitude, exercise, nature, play, and sociality do seem to impact levels of wellbeing and life satisfaction.[454]

Trying to control our thoughts or suppress aversive thoughts is not possible and can be corrosive. In fact, these things should be avoided. Attempts to suppress aversive thoughts paradoxically amplifies them, as well as the TR. Our thoughts, ideations, and constructs deserve expression, analysis, and refinement. We must concede they are a part of our physiology that is determined by the interface of the exposome and the genome—things not entirely within our control. We must be cautious about reacting and clinging to our thoughts, as they may not reflect reality. This expression, analysis, and refinement are easier to do when we are feeling safe and flexible, and can detach, rather than when we are feeling threatened, rigid, and holding on.

As an example, take a moment to deconstruct a fundamental cultural construct within the United States—"laziness."

- We know that when threatened, asocial, antisocial, and maladaptive behaviors arise. These maladaptive behaviors not only make all of us unsafe, but they also cost all of us significant energy, resources, and money—they are expensive.
- We also know when people have basic financial security, they are healthier, work more and produce more—they do not become "lazy."[455] Safety physiology is not only prosocial, but also pro-work, and conducive to production.
- More notably, T2 physiology, characterized by low metabolic rate, fatigue, immobilization, and isolation, makes work and production very difficult.
- On examination, "laziness" is a false construct and false narrative driven from a state of fear and anxiety in the descriptor. Within this narrative laziness is described as a characterologic defect in another human, to perhaps

motivate them through social threat to do more. Or it may be to marginalize them. Or perhaps it is meant to degrade them for the descriptor's own self-image needs.

- Inactivity in individuals in threat is protective. Inactivity in individuals in safety is restorative. The construct of "laziness" is neither protective nor restorative—it is destructive. We repudiate the "lazy," thinking it will change behavior when it only serves to reinforce T2 inactivity.
- Not only is lifting someone up who is stuck in a pit of despair anathema in our culture, but so is acting to prevent them from falling into this hole. In fact, we dig these holes and kick people into them, resulting in their demise—as well as our own. We then complain about the holes and the people in them.
- Based on the evidence available to us, this is very irrational. People in threat do not think rationally; we are a culture in threat.
- There is then an expectation for those within the pits of despair to "pick themselves up by their bootstraps"—a physically impossible task.

There is no way up in this scenario; the expected is impossible. The adherence to this false construct is insanity. When we feel safe and flexible, even curious, it is easy to deconstruct these false constructs, narratives, and beliefs. Afterward it is time to reconstruct. It is time for a new paradigm that will also be subject to expression, analysis, and refinement—no clinging.

(A thought on thoughts: Think of thoughts as soap bubbles, slippery, fragile, temporary, to be played with, and finally to let float away and pop into nothingness.)

Just as our thoughts aren't controllable, in some regards, the same applies to our behaviors. Physiology can determine action prior to our awareness, let alone a thought that determines an action. Our physiology can move us, and then our cortical and cognitive awareness and thoughts, ideations, constructs, and narratives catch up to the action. A reflexive narrative can be formulated after the behavior has been initiated and prior to cognition especially in times of threat.

Two roads diverged in a yellow wood,
And sorry I could not travel both
And be one traveler, long I stood
And looked down one as far as I could
To where it bent in the undergrowth;
Then took the other, as just as fair,
And having perhaps the better claim,
Because it was grassy and wanted wear;
Though as for that the passing there
Had worn them really about the same,
And both that morning equally lay
In leaves no step had trodden black.
Oh, I kept the first for another day!
Yet knowing how way leads on to way,
I doubted if I should ever come back.
I shall be telling this with a sigh
Somewhere ages and ages hence:
Two roads diverged in a wood, and I—
I took the one less traveled by,
And that has made all the difference.

Robert Frost

Note: Split-brain analysis, where the right and left hemispheres of the brain have been disconnected from each other (for medical reasons such as controlling seizures), demonstrates that adopted narratives generated within the left cerebral hemisphere can be complete confabulations in relation to the real reason and intention of what the right brain set out to do. Yet, we hold to the belief in the narrative with certainty the adopted narrative is correct.[456] Random thought bubbles and external inputs flow through the neocortex to reach the cognitive centers of the brain to be formatted into constructs—real or not, to be transferred to language centers for narrative formation—real or not, to be rehearsed and embedded in memory and belief—real or not, to be certain of, cling to, fight over, and kill for. Belief is simply an opioid for the human neocortical awareness and knowledge of our fragility and the horrific possibility of our annihilation. That should give us all pause.

(I have typically sequenced this prosess as thought -> created ideation -> cognitive construct -> symbolic narrative -> held belief, but arguably when in threat a reactive narrative can commence prior to cognitive construction, whereas in safety a narrative is more likely to follow cognitive construction. Regardless, whether an automated narrative or a cognitive narrative comes first or second, both can be pushed into a held belief.)

Trying to modify behavior through mental functions can be problematic, particularly when we are in threat and predisposed to reaction while void of contemplation. Yet, when we can choose, we should choose safe, adaptive behaviors, as they drive safety physiology and reinforce the physiology of healing, health, wellness, and happiness. There is bidirectionality in this process. Physiology determines thoughts and behaviors, yet thoughts and behaviors can influence physiology to some degree. The neocortex does have a role in modifying thoughts and behaviors. Contracts, ideations, constructs, narratives, and beliefs are not solely determined from below, and can be created from above, but more so when safe versus threatened. In this regard, safety can compound safety just as threat can compound threat...and both counter each other.

Information does not in and of itself do a very good job of changing emotions, behaviors, nor thoughts. Neither does misinformation. The content of information is less influential than the context of threat versus safety and the associated physiology and emotions surrounding the information. Information and misinformation can be part of a platform that can shift social contracts, created ideations, cognitive constructs, symbolic narratives, held beliefs and pervading culture, but what information we attend to and integrate is very much a systemic determinant connected to our physiology. Because of a bias towards threat, information that is threatening can be sticky and prioritized. However, the data itself may not change behavior, but the physiology associated with threat versus safety very much may.

"Wine is a mocker, strong drink is raging,
and whoever is deceived by it is not wise."

Proverbs 20:1

Alcohol accounts for over 150,000 deaths in the U.S. annually. It is a leading cause of death in the U.S. just behind tobacco, poor diet, and inactivity. Alcohol associated deaths in the U.S. rose 30% between 2016 and 2021. Alcohol, a toxin, stimulates a TR. Used chronically, it can increase all the associated diseases related to chronic threat. Alcohol is one of the leading causes of infidelity at both a social and cellular level. Alcohol is a leading cause of asocial and antisocial interpersonal and intercellular behavior. Alcohol is a leading cause of cancer and is classified as a Group-1 carcinogen according to the National Cancer Institute. Yet, this information does relatively little to change behavior across the population.

Why?

Alcohol, despite its toxic nature, dissociates and this feels good. In addition, drinking alcohol is associated with fun social activities. Alcohol has been entrenched into culture for millennia. It has been systematically marketed to appeal to the constructed false image of the self. Deception and manipulation, targeting emotions and image, are hallmarks of marketing, under the guise of sharing information. Marketing is inherently predatory and designed to take advantage of our neurophysiology—humans want to feel and look good, **and** fit in. Alcohol consumption is perceived and believed to prevent a social death. Finally, alcohol is an addicting toxin—it is hard to quit.

Humans need to resolve conflicts between incongruous facts and behavior. Thought suppression, emotional repression and denial are common, as well as threat associative false narratives and rigid beliefs. But facts and physiology do not deceive, manipulate, or lie. They expose truth…for us to ignore.

"It is not the strongest of the species that survives, not the most intelligent that survives. It is the one that is the most adaptable to change."

Charles Darwin

We have walked up an evolutionary tree from the level of the code for life itself—our genome—found within the nucleus of our cells, to the cells themselves, to the primitive autonomic brain, to the emotional brain, to the more advanced social brain, to the most advanced mental brain. We have addressed what it means to be safe at each of these layers. Next, we will step away from the individual organism to the collective—the colony.

Societal Safety

Safety and security don't just happen, they are the result of collective consensus and public investment. We owe our children, the most vulnerable citizens in our society, a life free of violence and fear.

Nelson Mandela

The United States Constitution does establish a foundation for the safety of the individual and the whole. The Constitution, even within its short preamble, is an expression of the tenants of TVST—ensure safety, defend threat, promote welfare, secure freedom, as well as protect and propagate the species.

> *We the People of the United States, in Order to form a more perfect Union, establish* ***Justice****, insure domestic* ***Tranquility****, provide for the common* ***Defence****, promote the general* ***Welfare****, and secure the Blessings of* ***Liberty*** *to ourselves and our* ***Posterity****, do ordain and establish this Constitution for the United States of America.*

The Constitution establishes boundaries and jurisdiction within the federal government and provides a balance of powers that requires adaptation, cooperation, collaboration, and compromise for its success. The Constitution also defines the boundaries and jurisdiction between federal and local governments.

Within the Amendments, most specifically the Bill of Rights, individual boundaries and rights are defined and protected. These address the right to expression, property, self-sovereignty, and self-defense.

In addition, the Constitution describes the role of government to defend against threats and support the safety of the general population fairly, equally, and with equity. Welfare is not a dirty word within the Constitution; it is a requirement.

The United States Declaration of Independence and the Constitution have a clear contrast. The former in combination with the Bill of Rights sanctifies the rights of the individual, and the latter, the importance of the common good. Lean one way and individual wealth and disparity unbalance the scale. Lean the other way and governmental bloat and waste unbalance the scale. It isn't an either-or situation. Both are needed. To survive we need multiple phenotypes in biology and sociology. We need a strong democracy that contains elements of both rugged individualism with competition and economic capitalism and empathetic societalism with cooperation and economic socialism to be used to each of their strengths, and to be flexible and adaptable to the needs of the time. No band, tribe, village, city, or country can be held together and exist without societalistic functions especially for accountability and defense purposes, but also the general welfare.

Given the times at the creation of our founding documents and the profound fears of monarchies and autocracies, perhaps our founding fathers biased America towards rugged individualism and away from centralized collectivism. There was much deliberation and uncertainty in the creation of these documents, conceding progress over perfection, and with the idea that the Constitution wasn't to be a static document but a living organism in search of maximizing safety for the individual and the whole, these documents were signed and institutionalized. As the world changes, strict constitutionalism should reflect the changing times and this desire to provide safety for the citizens of the U.S., not the written word or lack thereof.

There is no ideology in biology. The ever-changing exposome meets the genome and phenotype changes to address the needs of the moment. Governance should be similarly flexible, void of ideology, but full of ideas and debate. We should be sharpening the edges of capitalism and socialism to be used as precise tools of transcription, not conducting warfare on each to blunt their effectiveness. We frequently describe a tug of war between individual rights and the common good, where one side must defeat the other side. A binary win-lose proposition. We should see this not as a battle, but as a balance, absent the metaphors of conflict and war, yet composed of the realities and benefits of opposition and competition, and more so of cooperation and adaptation.

"As soon as the land of any country has all become private property, the landlords, like all other men, love to reap where they never sowed, and demand a rent even for its natural produce."

Adam Smith

We should also acknowledge that too often the scale is weighted by and towards individual, old, straight, educated white men, while systematically denied to others. Currently, three older white males in the United States own more collective wealth than the bottom 50% of the population combined, and 1% of the population holds the majority of the total wealth in the U.S. This cumulative wealth is approaching $50 trillion dollars—with a T! Morbid wealth and morbid poverty are the grotesque inequities of autocracies and monarchies that our founding fathers and our founding documents set out to avoid...that we have desecrated.

"They never invited the Plain-Belly Sneetches – they left them out cold in the dark of the beaches."

The Sneetches – Dr Seuss

The self(selves) is a polymorphic defense state, a degenerative state, but a necessary state for survival. The static-self is illusory. Within this illusion, the ideation, construct, narrative, and belief that we are all static distinct individuals is at the core of threat related behaviors and maladaptive outcomes including income inequality, racism, climate change, cults of personality, and just about every other source of human misery. Our focus on the individual within a health care model is therefore suspect. A focus on population safety and health with individual incentives for safe and healthy behaviors should have far better outcomes and cost much less money than what we are doing today.

It should be reiterated that as safety, not wealth, leads to wellness, health, happiness, and thriving, we cannot expect to achieve wellness, health, happiness, and thriving by pursuing mutagens, allergens, antigens, pathogens, structures, and wealth, nor star bellies, within a world full of threat and a vacuum of safety. Small-scale, nonindustrialized societies of indigenous peoples have high levels of life satisfaction, comparable to those of wealthy countries, despite very low monetary incomes. Human societies can support very satisfying lives for their members without necessarily requiring high degrees of conflict, competition, discrimination, production, accumulation, and monetary wealth.[457]

We must treat at the generational, historical, ecological, physical, emotional, social, mental, societal, cultural, financial, and spiritual levels to truly thrive in this world. Nothing in isolation trumps the pursuit of safety at all these levels for achieving health and wellness. Any treatment algorithm that falls short of this will likely fail. This will require both individual, as well as population-based, interventions.

Cultural Safety

> *"Culture precedes positive results. It doesn't get tacked on as an afterthought on your way to the victory stand."*
>
> *Bill Walsh*

The happiest and wisest countries in the world recognize the balance between the needs and strengths of the individual versus the needs and strengths of the colony.[458] The Scandinavian countries have some of the highest happiness and health levels across the globe. Their strong social welfare system reduces threat and promotes safety, while their well-regulated market system produces and allocates resources fairly and efficiently. They have found a balance between individualism with economic capitalism and societalism with economic socialism that plays to the strength of both, and thus promotes the welfare of both the individual and society. This is not only reflected in their health and happiness levels, but also in other safety traits such as trust, empathy, compassion, kindness, and low crime and unemployment rates. Interestingly, Japan ranks significantly lower than the Scandinavian countries on the happiness scale, but is number one on safety, health, and longevity scales. This perhaps indicates that health and happiness parallel each other but are not the same thing.

According to the World Happiness Survey, Finland has ranked as the happiest country in the world for the last seven years.[459] Correlates include lower income inequality (the difference between the highest paid and the lowest paid), high social support, freedom to make decisions, and low levels of corruption.

As noted previously, threat can be relative to those who have less compared to those who have more. Those who have less may have real threat. However, sometimes it is just relative threat physiologically and emotionally represented by envy and jealousy. Sensing relative threat or unfairness is rooted in evolutionary physiology with the goal to protect and propagate the genetic code. Fairness and sharing of resources allow for healthier organisms and species to help sustain and propagate their code. A dominant and greedy organism operating autonomously and to the detriment of the other organisms and the species is very much like an aggressive cancer in some sense. Biology, well preceding any neocortical functions such as social contracts, created ideations, cognitive constructs, symbolic narratives, or held beliefs, let alone concepts of ethics and morality, directs organisms to eliminate the threat of infidelity to the whole. Humans, too, have some primitive biologic drives to take down those who have taken too much, or simply have too much.

The affluent biologically and physiologically become unsympathetic targets of the others within the species. In addition, those who have more have more to lose. This combination can make the wealthy fearful and protective and thus exhibit all the physiology, emotions, behaviors, thoughts, ideations, constructs, narratives, and beliefs consistent with chronic threat. In addition, the affluent fortress and isolate for their protection and the protection of their stuff and their money. They focus on individual survival, not on the species' goals to protect and propagate the code.

We need to recognize that both extreme and morbid wealth and extreme and morbid poverty can have adverse consequences for both individuals and societies. Inequity exacerbates and compounds threat on both sides of this equation. Balancing and minimizing inequities can result in increased resources available for the health care, education, and welfare of a society. Equity converts to happiness and health. On the other hand, inequity, even if it is only relative threat, can lead to conflict. History tells us inequity is metastatic and kills.

However, caution should be used when invoking economic socialism as a primary or solitary treatment for inequity. This may be no different than the treatment of the symptoms of illness, disease, disability, pain, and suffering without ever treating the root cause of these symptoms—threat. After all, Robin Hood was still a criminal. However, what may be more criminal is the acquisition of monies to be squandered in inefficient and ineffective societal programs with the illusion of doing good while reinforcing disability, pain, and suffering. Too frequently we apply a salve to the sores and symptoms of inequity without exposing and treating the root of the infection, only to see the problem fester, if not worsen.

Inequity is both a symptom for and a compounding factor of threat. Beyond the redistribution of resources, the total TL must be addressed to restore safety, health, and happiness. Rebalancing these scales through financial means is only a part of the total equation. Population-based needs and safety strategies will work much better. Additionally, whether using economic capitalism or socialism for restoring equity, programs require transparency, accountability, and continuous quality improvement to ensure efficiency and efficacy. Missing these things both can enable dysfunction and destabilize societies leading to more disease, disability, pain, and suffering.

The biology of culture is significant. A culture filled with resentment, competition, conflict, and threat will breed illness, disease, disability, pain, and suffering. A culture designed to reduce threat and provide safety leads to wellness, health, and happiness. The focus on minimizing cultural threats and maximizing cultural safety should in and of itself reduce inequity, disenfranchisement, prejudice, and injustice without the need for forcibly mandating, legislating, and enforcing equality through government initiated wealth transfers.

Governmental intervention without a shift in cultural ideations, constructs, narratives, and beliefs, plus consensus, can become another source of conflict and threat. That is not to say policies, procedures, and laws are not necessary to improving our lot, but building a culture of safety within people more than institutions, must be undertaken in pursuit of wellness, health, and happiness primarily, but not exclusively.

Culture can be built outside of these mechanisms for policies, procedures, and laws. People and leadership are important. This leadership is not of vision or supervision. It is not formed from rhetoric but from example. It is not about having someone carry your water for you, but about you carrying someone else's water.

Leadership is about humility, mentoring, teaching, and

sacrificing. It certainly is not the worship of the individual or an individual, but the worship of the whole. There will be leaders and followers, but in well-run organizations leadership is diffuse, and it is hard to see who is in charge. Rulers and subjects, nor autocrats and sycophants, should not be the order.

"If I then, your Lord and Teacher, have washed your feet, you also ought to wash one another's feet. For I have given you an example, that you should do as I have done to you."

John 13:14–17

To return to Japan, their culture and society are extremely safe, but contain more social contracts, constructs, narratives, and beliefs than many cultures and societies that rank higher on a happiness index. In Japan, the sense of emotional and behavioral freedom may be more restrictive. Safety and freedom can coexist, but they are different and may weigh differently in the determinants of health and happiness. This may explain relatively lower happiness scores relative to health and longevity in Japan.

Culture-bound disorder refers to a classification of emotional, behavioral, social, and mental disorders or syndromes that are considered specific or closely related to cultural factors. In Japan, two culture-bound disorders, *taijin kyofusho* and *hikikomori*, have been described. The first disorder, taijin kyofusho, is a culture-bound social anxiety disorder. It is characterized by the experience of anxiety in, and often avoidance of, social and interpersonal situations for fear of offending others through blushing, eye-to-eye contact, body deformity, and/or emitting body odor. The second disorder, hikikomori, is distinguished by symptoms of social withdrawal, self-confinement in one's home, no intimate relationships with family members, and the absence of engagement in social activities.[460] These two described disorders closely resemble the T1 and T2 phenotypes respectively.

There is speculation that because of the profound role the construct of shame plays in Japanese history and culture that these disorders have shame at their root. TVST suggests that at the root of these disorders is the threat of a social death, brought on by false or maladaptive cultural contracts, ideations, constructs, narratives, and beliefs.

These cultural observations and the power of the threat of a social death thus raise concerns regarding broadcast and social media that create a culture that demands conformity versus risking rejection. Threat culture births toxic cults. Threat culture is a formula for declining wellness, health, and happiness. The broadcast and social media platforms have the potential to increase connection, expression, freedom, and safety in the world. But unfortunately, they can also be weaponized to increase deception, manipulation, and threat in the world. The depersonalization of communication through social media, the loss of face-to-face interaction, reduces empathy, compassion, and kindness. This is fertile ground for bullying and antisocial narratives and attacks.

The marketing and profit strategies of broadcast and social media can set culture and prey on susceptible human biology. This includes the vast biology below the level of human cortical awareness that is engaged in our search to find and then avoid threats—particularly our fears of missing out, being left out, or being kicked out. It should be clarified here that this concern with media is not inclusive of the broadcasting of information or of entertainment but is specific to the broadcasting of misinformation and manipulation.

Overt and covert cultural, social, and emotional threat strategies can be used to manipulate humans for power, prestige, and profit—despicable, but true. Power, prestige, and profit come at the expense of the individual's, society's, and culture's health and wellness. This coercion smells of fraud in that it is deceptive and insidious in its use of biologic strategies, and even computer-generated algorithms. Like environmental toxins and pollutants, these advantages come with both overt and covert costs to individuals and societies.

The hidden cost of the deception and manipulation typically comes at the expense of the targeted for the unjust benefit of another individual, their party, their company, or their country. Accountability for any asocial, antisocial, or antisocietal behaviors is essential, whether these are individual, corporate, or governmental transgressions.

Connection and information are decidedly different than deception and manipulation. Deception and manipulation are serpent sneaky and almost always harmful. Aspects of marketing, campaigning, media, and AI are metaphorically running hand in hand with Satan, the Devil, the great trickster in how they alter human contracts, ideations, constructs, narratives, and beliefs from deep below cortical consciousness. We have no idea, and it all seems so real. Technologies play a significant role in the mental health crises of today (particularly with regards to social media's effects on our youth).[461] Both external and internal threat algorithms drive us to consume excessive and unhealthy foods, goods, and content.

Politics has become vapid media with the adoption of marketing strategies empty of any substance. It is now a persistent pursuit of prestige, power, and profit by any means possible, but to no meaningful end. The demonization of the other within politics amplifies threat and threat associated behaviors and is counterproductive to the goal of life itself—to protect and propagate the human genetic code, let alone longevity, health, and happiness.

Just as free markets need transparency to be healthy, so do broadcast and social media. Both demand inspection and regulation—illumination is an excellent disinfectant. These people, parties, companies, and some nations, too, are not paying full costs for their acquired power, prestige, and profit. Their power, prestige, and profit come at a great expense to other individuals and societies. Marketing and the algorithms used in social media, are the toxic pollutants of the 21st century that will require control, detoxification, legislation, inspection, regulation, prosecution, and restitution to be cleaned up adequately. Deception and manipulation for power, prestige, and profit is not the same as freedom of opinion and speech.

Social media's neuropsychological algorithms are designed to addict to social media. Under the disguise of social connection, the connection on social media is to the algorithm, not another person. Social media is asocial if not antisocial. In a perverse twist the best solution for this "social" media addiction is true safe social connection with other people and the world without a technological interface.

It is fair to say that boundaryless (marketing algorithms and social medial are surreptitious in their invasiveness) social contracts, paranoid delusional ideations, faulty cognitive constructs, false threatening narratives, and rigid mandatory beliefs do not constitute culture but constitute a cult. A cult can prove difficult to legislate, regulate, and enforce against. It requires forming and managing the proper culture—a culture of safety—to push back against these forces.

"Diversity doesn't look like anyone.
It looks like everyone."

Karen Draper

Diversity of genome, ideas, and people are important to a healthy society and culture. Rigidity of thought and fear of different people are hallmarks of a TR. Consider that modern humans have now overpopulated the Earth while Neanderthals died out approximately 40,000 years ago. This is frequently attributed to Homo sapiens having better brains and being smarter, but the reality is Neanderthals had bigger brains, were likely equally creative and smart, and were bigger, stronger, and better hunters than Homo sapiens.

What Neanderthals lacked was diversity. Genetic studies have demonstrated that Neanderthals lived within small, closed bands, were inbred, and lacked genetic diversity and flexibility. By contrast, Homo sapien genetics were/are more diverse and flexible. Human societies also reflected this diversity, growing larger from bands, to tribes, to cities, to nations, and to alliances. Homo sapiens' large social networks added protective diversity and flexibility of genetics, as well as diversity and flexibility of thought.

Having opportunities to share genetic information increase genetic variability and flexibility to be better able to fight off pathogens and predators. Bacteria who share code, let's say for antibiotic resistance, enhance their chances for survival. In addition, a flexible genome can express many phenotypes (infinite?) through a multitude of epigenetic modifications to better meet the needs of any moment—also an excellent survival strategy.

Additionally, Homo sapiens, because of their larger communities, have a large cumulative IQ and historically have been good at cooperating, collaborating, adapting, and creating by combining their brain power for the advantage of the group.

Threat decreases these qualities and assets. This is something to contemplate as we see cults obstruct and fight diversity and idealize and fight for supremacy—perhaps to their and everyone's suffering and demise.

TBD.

Threat changes phenotype, physiology, and the connectotype within the brain. Threat brings out bad social contracts, paranoid ideations, faulty constructs, false narratives, rigid beliefs, and shifts culture. When culture accepts threat- biased beliefs as reality, then a form of group psychosis becomes what is "normal"—at least within the cult. We must accept that under chronic threat, we all suffer from at least a little bit of schizophrenia. Schizophrenia is a symptom of a phenotype, not a pathotype, not an illness in the usual sense, but a phenotype of chronic threat. It is this slippery slope we are on.

Spiritual Safety

> *"Heaven is long-enduring and earth continues long. The reason why heaven and earth are able to endure and continue thus long is that they do not live of, or for, themselves."*
>
> *Lao Tzu*

There is still an additional space to be addressed, that being the spiritual. The spiritual space does not follow evolutionary progression. In fact, the spiritual space encompasses all. The spiritual space leaves behind constraint, repression, suppression, contracts, ideation, construct, narrative, and belief—the human neocortex—to discard the self for a deep connection to the world around us. If social connection is defined as the relationships between individuals of the same species, then spiritual connection is the connection to the rest of the cosmos beyond the species.

Life at its very core is about connections and relationships—life arises at the interface of the exposome and the genome. Connection is inclusive of human-human relationships, but not exclusive of other relationships.

Spiritual connection and safety can be found within deep non-Homo sapien relationships. These connections ease our existential angst, and the threat and fear of annihilation. Faith displaces belief. The "sapiocortex" spins thoughts into created ideations, cognitive constructs, symbolic narratives, and rigid beliefs. The collective can reenforce this process, but ultimately this is a very internalized process. Whereas the spiritual realm is oriented externally with a detachment from this internal world and the self to drift into a world of expansiveness, curiosity, awe, and faith. One state reflects a threat phenotype while the other a safety phenotype.

"Everything in excess
is opposed to nature."

Hippocrates

Spiritual connection can most easily be found when in nature. In nature TS and TC drop, cortisol levels normalize, parasympathetic tone improves, we are more receptive to other people, and our TR disappears. Research has shown that adolescents raised exclusively in rural environments have larger hippocampal volumes and better spatial processing than children exclusively raised in cities. Other research demonstrated that spending just an hour in nature leads to a decline in amygdala activity in adults, whereas it remained stable after walking in an distressful and busy urban setting. Proximity to green or blue natural waters is associated with less anxiety and depression. Every additional 360 meters from the nearest green or blue space is associated with a 10% greater risk for anxiety and depression. Forests improve neocortical functions, reduce amygdala TR activity, and increase overall wellness and health. The Japanese participate in forest bathing (Shinrin-yoku) to improve their health and happiness.

Nature has a pulse and a rhythm. The smells, sounds, and

sights in nature create a calming outward focus. Time in nature decreases impulsivity, anxiety, rumination, obsession, depression, obesity, cardiovascular disease, and increases attention, concentration, memory, quality of sleep, and a sense of well-being. Nature causes a shift away from the DMN and self-obsession. Nature opens the door for awe.

The opportunity for these connections is all around us and available all the time. It is especially abundant and clear when we are safe and open to this level of connection. Spiritual connection is a synchronization and resonance of our cells, tissues, and organs—our being—with other beings and the surrounding environment. The colors, sights, sounds, smells, and molecules of nature, including the molecules within the air and the soil of a forest, lower TS and TC. This creates a deep connection that converts to health and wellness. Spiritual connection requires time and space. It is a place of absorption, acceptance, calm, and at times, awe, that is profoundly healing.

Feelings of awe shift our focus from the self to the other and connectedness. Awe defeats narcissism and entitlement—threat states. Awe brings forth the biologic truth of our connection and interdependence. Being in nature and experiencing its expanse, complexity, and beauty is one of the best ways to experience awe.

Another way to invoke awe is the observation or contemplation of the moral beauty of another. The courageous, compassionate, kind, and prosocial qualities of the human species induce awe in us. As if a contagion, these qualities lead to more prosocial behaviors, healing, health, wellness, and happiness.[462]

We have a crisis of threat. We tend to double down on the crises of threat with self-focused solutions to unhappiness, disease, disability, pain, and suffering. We increase threat physiology in these processes. We make people sicker. The pursuit of individual pleasure and happiness through the concepts of the authentic-self or true-self or through the concepts of self-care, self-help, self-love, self-compassion, self-awareness, self-examination, self-realization, self-actualization, self-image, self-esteem, self-improvement, self-….is a false paradigm that holds us to an illusion of the self and the defense of this illusion. This holds us in a TR. It is time for a new paradigm that orients us to the outer world, deeper connections, our interdependence, and away from the defensiveness of the false-self. We need a new paradigm for the protection and propagation of our Homo sapien species, and the other species on our Earth.

We are missing the spiritual in the world today. The spiritual that was so much a part of ancient cultures has been marginalized in the modern world. The spiritual needs resurrection within the modern world. Ancient precivilization native cultures may prove to be our guides back to wholeness.

We need to place boundaries around work, media, screens, and indoor time, and prioritize time outside in nature to contemplate, meditate, play, labor, and socialize.

Individuals can adopt the strategies of wellness and health that are applicable to the individual to afford healing. However, for any and all of us to truly heal, be well, healthy, and happy, we will need institutional, societal, and cultural change. The spiritual and safety should guide these processes, too. The spiritual must be prioritized and sacred.

More Good, Please

> *"Dualism is an isolating sense of separation, a feeling of being fully on one's own, unconnected to others, to the natural world, to the whole of reality. Overcoming dualism is the cure for suffering."*
>
> *Dale Wright*

The wave rises and rushes to crash upon the shore to then recede back into the ocean. The wave has its time, form and force, but its contents remain the ocean. As the wave rises, rushes, crashes, and recedes its contents are ever changing, but still the same, still the ocean. The wave is always connected, continuous, and confluent with something greater as a part of the vast ocean, never separate from the ocean, yet observable and measurable in its time, form and force.

TVST purposes that life is similar in this respect. Lifeforms arise and rush forward to eventually crash and recede back into a universal ocean of consciousness. Life manifests consciousness in this way and consciousness is the source and the protector for life. Consciousness is the ultimate connector for life, but consciousness also gives the awareness of separateness as without this awareness life would not protect itself and the organic would quickly become the inorganic. Life is ever changing, yet never separate from the field of consciousness. Despite this profound interconnection for humans the field of consciousness is not fully perceivable nor measurable...for now? But its presence is apparent within lifeforms. The problem here is that humans are unable to see the connective, continuous, and confluent nature of lifeforms and particularly this relationship to humans, including the relationship of a human to another human.

This failure of human perception leads to humans only being aware of their separateness and not their connectedness. TVST has purposed that under conditions of threat, this perception of separateness and the lack of connectedness only gets stronger. This biased perception towards separateness leads to both social and existential angst, and eventually to biased and then false ideations, constructs, narratives, and beliefs. These false ideations, constructs, narratives, and beliefs support the creation of the false-self, but also the false-other. Under threat, we tend to create "positive" notions of ourselves while creating "negative" notions of others...to further disconnect us. Asocial and antisocial emotions, behaviors, and thoughts arise within states of threat. Safety helps to restore all connections and sociality.

"An eye for an eye will leave
the whole world will be blind."

Mahatma Ghandi

In times of severe chronic threat, we tend towards immobility and inaction in form and force, and rigidity and inflexibility within thought. We can get stuck in a "sapiocortical" rut and fail to take action to actually resolve the TL. Instead, this leads us down the path of investing in and clinging to threat biased thoughts and false ideations, constructs, narratives, and beliefs to give us the illusion of control and the illusion of safety, when we would be better off focusing on what we need to do to actually become safe. This threat path can ultimately lead to ideations, constructs, narratives, and beliefs around God, and ultimately the formation of societies and cultures around this process and its illusions, i.e. religions. Upon this fearful, fragile, and faulty foundation we attempt to define God.

Just as when we observe the dynamics of space-time we change the dynamics of space-time, when we define God we change God—we define a false-God within an illusionary human-centric design to try to keep us safe, or at least feel safe. However, this "sapiocortex" generated false-God and the attached ideations, constructs, narratives, beliefs, and religions become paradoxical with regards to safety. Connection, the perception of connection, and safety are lost as humanity divides into idolatrous ideologic driven factions that escalate competition and conflict within the world. This is fertile ground for demagogues.

The vastness of connection becomes the constriction of disconnection, good becomes evil, peace becomes war, joy becomes suffering, wellness becomes illness, health becomes disease, life becomes death, safety becomes threat, and the creator becomes the destroyer. This false-God morphs into the Great

Trickster as the desire for The Good and to be good perversely slithers and twists itself into evil.

The reality is this isn't the competition between evil and good, but the consequence of threat versus safety.

This paradox makes it most pragmatic to concede that humans cannot define God in an ideation, construct, narrative, belief, or religion. God is well beyond our perception and any notions we might concoct. Any attempt to know or define God will at best fall short. At worst, it will create more dysfunction, disease, disability, pain, suffering, and death in the world. Just as empathy cannot be taught but must be felt, God cannot be taught but must be felt. The Good and its ultimate connection to all including the universal field of consciousness is found when we are safe. It is best to remain agnostic and silent of and on God. Believe not of God, nor speak of God, but have faith, and seek and provide safety, to feel The Good.

"To one who has faith, no explanation is necessary.
To one without faith, no explanation is possible."

Thomas Aquinas

Be wary of those who speak and claim to know of God. Knowingly or unknowingly they are working within a paradox to ultimately do more harm than good.

Temples, mosques, synagogues, churches, and cathedrals should be places of safety for nurturing both S1 and S2 phenotypes—spaces of connection, respite, solitude, prayer, and meditations, as well as compassion, communion, celebration, feasts, music, dance, play, and service. Bastions of safety, but not places for the human ideations, constructs, narratives, beliefs, nor the unknowable word of God. They should be places beyond human "sapiocortical" knowledge and intellect, but more so beyond human "sapiocortical" illusion, manipulation, and deception. They should simply be places for all humans to be safe and to feel The Good.

Threat biased false ideations, constructs, narratives, and beliefs blind us to our confluence, continuity, and connections. Faith, not knowing, will make us and the world safe and well.

"To have faith is to trust yourself to the water. When you swim you don't grab hold of the water, because you will sink and drown. Instead, you relax, and float."

Alan Watts

Accept the Flow—but Embrace Solutions of Safety

"Those who flow as life flows know they need no other force."

Lao Tzu

Remember life is a flow between the phenotypes of threat and the phenotypes of safety. Just as with individualism and societalism, capitalism and socialism, or competition and cooperation, it is not an either-or, but both. Life is not to be and cannot be balanced in static equilibrium. Life contains pain and life should contain joy. Life rejects utopia. However, the time spent in safety must be plentiful. Hopefully, at a minimum, time spent in safety will match the time spent in threat, but preferentially, exceed the time spent in threat. This will allow us to become healthy and happy; to grow, regenerate, and live long; to be resilient, to reproduce, and to successfully propagate the code.

Our ideations, constructs, narratives, and beliefs can block a more natural flow of life. In life we will ride some ups and downs, but we don't need to concoct more downs.

We tend to create problematic narratives full of moral platitudes to be inspirations and motivations towards a successful life. In our consciousness these can feel so obvious and right, yet below consciousness the unobvious plays out within a physiology of conflict and threat to be oh-so wrong.

- Be boundless—shoot for the stars.
- Anything is possible—all you have to do is believe in yourself.
- You can be anything if you just apply yourself—you can even be president!
- Either you fight or you give in.
- We have to take the fight to them.
- I will fight for the American people.
- Winners are losers who got up off the ground and gave it one more try.
- Set your goals high, and don't stop until you achieve them.

These narratives are to dream big and be willing to struggle and fight to achieve these dreams. Many people believe that to have a successful life one must construct an aspiration and a desire, set a goal with a high bar, and relentlessly and aggressively pursue the construct to satisfy all the accompanying cultural platitudes. This process is believed to give a sense of meaning and purpose to life that will ultimately lead to success and fulfillment.

Besides human beings, what other life form does this? None. Then how can this be the right formula for life?

Life is hard—for all life forms. Life is full of threats. Defenses are required. Life is at times a struggle, and sometimes even a fight, to keep the organic from becoming the inorganic. All life eventually dies. The only known real purpose of life is to protect and propagate the genetic code of life—the purpose of life is to successfully sustain life through the generations. Life is not about the me, but about the we, and particularly the very wee—our children. The rest is platitude.

So why make life harder? What is so essential about pursuing a narcissistic battle for platitudes? A four-billion-year history of other life forms on Earth suggests that such human ideas, constructs, narratives, and beliefs are inessential to life. Life is less about a story and all about experience. The moving through and the connecting to the environment comprise the experience of life. While meaning and purpose are given with life—protect and propagate the code. There is no mystery as to the meaning of life. The search for purpose is unnecessary.

What is essential for all life forms is safety. Safety provides the time and physiology for growth, regeneration, recovery, reproduction, and resilience. Wellness, health, and happiness are found within safety. Humans have no essential need to create a false self, seek an impression of meaning and purpose, nor engage in unnecessary competition or conflict to live a "good" life. These illusions of control within the metaphor of a battle lead to our many wars within ourselves and with each other, to much suffering.

The incessant victim-martyr-hero stories of our culture glorify the conflict, competition, struggle, and fight, and the individual superhero who survives the battle to fight another day while ignoring the background of destruction left in the wake of the war. We aggrandize the battle between good and evil—always feeling we are the good, or at least the justified. We gravitate to wars on poverty, drugs, crime, wokism, cancer, viruses, peoples, countries...the "evils" of the world without addressing the need for the creation of safety. The war between good and evil is an illusion, while the choice between threat and safety is our reality. Choose wisely. Wars fix nothing. Safety fixes everything.

So, when we hear a speech about struggling, fighting, and battling to achieve aspirations and goals, to thus find meaning and purpose, to ultimately become a successful and fulfilled human being, we should be skeptical.

This is not the flow. Life is about fluid emotions, motions, movements, and motivations within a flow not artificial rigid constructs, narratives, and beliefs to achieve heroic status through a subjective battle between good and evil within a constant struggle.

When inspired to get up and conquer the world and when encouraged to fight, fight, fight!...just don't.

There are things we don't need to do, things we don't need to place time and energy into, but what can we do? Perhaps stopping with our stories and our wars and starting with **the** story, the meaning and purpose of life, and the care of the globe? Accepting the flow doesn't mean not seeking solutions. Solutions flow. Solutions are liquid mixtures of uniform distribution that without constriction easily flow and disperse—as does safety.

In May 2023, the World Health Organization outlined economic reforms needed to improve global health. These reforms are targets for reducing threat and increasing safety across the globe.[463] The report, titled "Health for All," discusses health and well-being as an investment in a country's future economy and overall well-being. This includes the reformation of taxes on wealthy individuals and multinational corporations and debt relief for low-income countries during pandemics and natural disasters. The report notes governments must work to reduce inequality, not just between nations, but also within them. Fundamental to reducing the discrimination and violence that contribute to ill health is the reduction of poverty. World leaders must embrace the necessary economic reforms to reduce inequality and the associated afflictions of the world.

Income disparity, inequality, and inequity are growing across the globe, and particularly across the United States, the home front. In addition, our current paradigm of illness care has enormous waste within it that robs monies that could be used to correct these disparities and improve health and wellness.

"If the misery of the poor be caused not by the laws of nature, but by our institutions, great is our sin."

Charles Darwin

As a reminder, the U.S. now expends over $4.2 trillion in total spending and over $13,000 per capita annually on our individualized illness (more than health) care, far more than any other peer country, while our brain size, intellect, health, and longevity decline. Half of all U.S. deaths in the population under sixty-five would have to be avoided if the U.S. were to have mortality rates like its peer countries. In 2021, there were 26.4 million years of life lost due to excess U.S. mortality relative to peer nations. 49% of these lives lost died before age sixty-five. Black and Native Americans made up a disproportionate share of these excess U.S. deaths.[464]

Our current paradigm of fighting wars on various illnesses and diseases is not only failing us, but it is bankrupting us (more threat). Despite spending nearly three times more than any other peer country on "health" care, we have fallen to over 50th in the world in combined happiness, health, and life expectancy.

We do need economic and health care reform. All humans are entitled to safety. Safety saves money. Safety reduces the need for expensive health, environmental, and societal expenditures. Safety boosts productivity, the tax base, the stock market, and the gross domestic product (GDP). Safety reduces the threat of the national debt. Safety can resolve the ills of individuals, societies, and the planet. Safety compounds safety. When evolutionary biology and human physiology, instead of intuition and belief, dictate economic policy, we will thrive.

The determinant of chronic disease, disability, pain, and suffering is threat. Even relative threat, or inequity, is a cause of poor health and premature death. Equity is defined as the quality of being fair and impartial, but it is also defined as the value of shares issued by a company. The irony cannot be lost in that entitlement to corporate equity and private equity births societal inequity. Equity for the few is not equity. Inequity and inequality kill. Reform needs to focus not on the symptoms and correlates, but on the cause of individual, societal, and global illness, and disease. We must align resources properly, not waste them. We should dump "entitlement reform" in favor of "inequity reform."

Cumulative threat is the root of chronic disease, disability, pain, and suffering while safety is the spring that washes and soothes the inflamed, weary, and embattled to restore vigor, health, and happpiness. When the globe is inflamed, weary, and embattled perhaps it is time to give up the struggle and the fight for the mythical heroism and supremacy that we cling to with a dousing of safety. When inspired to get up and care for the world and when encouraged to serve, serve, serve!... just do...

"Nothing is softer or more flexible than water,
yet nothing can resist it."

Lao Tzu

Deconstruction

"Let Go or Be Dragged"

Zen Proverb

We have a variety of ways to terminate the life of a cell within our bodies. Some of these are programmed processes within the cells themselves while other deaths come from external processes to the cell.

Autophagy, a catabolic process, is somewhat the equivalent to recycling where cells are actively deconstructed, and the parts are repurposed for physiologic functions or the building of new cells. Recycling is built into our DNA and our cells.

Apoptosis is a form of programmed cell death described as altruistic death, or a sacrifice for the greater good of the organism. For example, the formation of certain tissues and organs may start disorganized. Subsequently, however, certain cells undergo active programmed dying to give a final form and function to these tissues and organs. Also, aging cells will undergo a programmed death. In their infidelity, cancer cells resist apoptosis, lose their altruism to the whole. There is no inflammation or gross destruction present within this form of death.

Necroptosis is an inflammatory form of programmed cell death. This is typically a process where the cell undergoes a strategic programmed death when apoptosis has been blocked, such as by a virus to sustain the cell functions for its own reproductive purposes.

Pyroptosis is an aggressive, inflammatory, lytic, messy, and less organized form of cell death in response to the infection of the cell by a pathogen. The goal is to kill the pathogen within the cell, but also to alarm the system of the threat and mount a systemic defense response to destroy the threat. Cellular death is a given and collateral damage is common, but biologically acceptable to kill to control a pathogen.

"To destroy is always the
first step in creation."

E.E Cummings

In addition, our immune cells can be summoned and act as a defense force to seek out damaged and abnormal cells to be destroyed and cleaned up, at times in a frank assault on rogue and antisocial cancer cells to kill them. This too can be accompanied by collateral damage to adjacent cells.

Biology outlines processes to deal with the no longer useful, the defective, the infected, and the antisocial at a cellular level. Deconstruction and repurposing are preferred, but sometimes the situation requires frank destruction.

The cellular deconstructive and recycling processes rejuvenate and reinvigorate the body and its physiology. They decrease threat and increase safety within the greater system. When these processes work well, we are healthy.

The cellular destructive processes are necessary for survival. However, this physiology makes us feel sick. If used chronically, ironically, it makes us more susceptible to further infection and illness, as well as cancers. These defense strategies that are toxic to pathogens and antisocial cells can be toxic to our own cells and activated relentlessly weaken us. The destructive processes can reach a tipping point into a downward spiral of chronic illness culminating in death.

The destructive processes are necessary for surviving, but the deconstructive processes not only help us in surviving but are part of our thriving. Deconstruction needs to come early prior to the onset of illness and disease. In deep threat states senescence surpasses quiescence. Senescent cells resist apoptosis and initiate and propagate diseases, including cancers.

This all begs the following question: if we need cellular deconstructive processes for our physiologic rejuvenation, health, and thriving, what else needs these processes? What else is obsolete, inefficient, ineffective, defective and/or diseased?

Our social contracts? Our created ideations and cognitive constructs? Our internal narratives and beliefs? Our egos? Our cultures? Our religions? Our policies? Our body politic? Our governments? Our institutions? Our cities? Our illusions of control? Our wars?

What within these is antisocial, creating more threat and not less, and detrimental to our health and happiness. Is cancer within the human body the equivalent to our cities within nature—why do they grow with a similar pattern? What is deserving of deconstruction and reconstruction, or even just destruction?

Our cells have these capabilities, but do we? We certainly need them.

Or more worrisome, have we passed the point of altruistic autophagy and apoptosis. Are necroptosis, pyroptosis, and cell to cell combat not only our present, but our condemned future?

Did we miss the window of deconstruction, repurposing, recycling, and sacrifice for our thriving to be marooned in a world of destruction, inflammation, aggression, and assault in a tenuous last-ditch effort for our surviving?

We can't afford to ignore these questions. Our certainty in faulty and failing beliefs, narratives, constructs, ideations, contracts, policies, laws, institutions, religions, and cultures may be our undoing. Failure to deconstruct tends to lead to destruction.

"War does not determine who is right
only who is left."

Bertrand Russell

As the genome builds the cell, our thoughts build contracts, ideations, constructs, narratives, beliefs, policies, laws, institutions, religions, and cultures. The genome, the program, the code, also guides the deconstruction and destruction of the cell. So too, we can deconstruct and destroy our failing and failed contracts, ideations, constructs, narratives, beliefs, policies, laws, institutions, religions, and cultures.

Death feeds life. In the cold dark quiescence of winter

leaves wither, let go, and fall to the ground to be recycled and repurposed to feed the forest. With the return of sunlight, illumination, warmth, growth, and lushness returns to the forest.

Our culture has become one of tradition, acquisition, and consumption. Biology suggests it should be one of deconstruction, sacrifice, and redistribution—a time not to hang on, but to let go.

Safety allows for flexibility, contemplation, and creativity to complete these rejuvenating processes.

Final Reflections

"The first problem for all of us, men and women, is not to learn, but to unlearn."

Gloria Steinem

As I reflect on my medical career, I recall being a third-year medical student on my first clinical rotation—hematology- oncology. Why start a medical student on hematology-oncology prior to having ever managed a blood pressure, an anemia, or even an abnormal electrolyte? The answer is manpower, or perhaps money. We were referred to as "scut monkeys"—dehumanized to workday and night, doing anything and everything requested of us expediently, efficiently, and enthusiastically for minimal reward. To remind us of our status, we carried the scut monkey pocket reference book filled with helpful tips on how to be a good monkey in our short white lab coats.

I remember sitting in the library late one night, perhaps early morning, feeling overwhelmed. I was searching for the latest and greatest chemotoxin protocol to try and afford a cure for breast cancer. Then, within a state of sheer exhaustion, I found myself trying to memorize the protocol to present on rounds the following morning. Alas, I was under the delusion that this protocol was the cure, yet in application it looked worse than the disease.

My suffering was nothing compared to the suffering of the patients. On an hourly basis, I observed the extreme suffering of cancer patients in that era. The cancer taking over their bodies was one threat just to be compounded by the infusion of various chemotoxins that threatened every cell in their being beyond the reaches of their cancer. They had withered muscles, lost hair, sluffed mucosa, painful ulcers, and horrendous opportunistic infections. They were nauseous and vomiting, pale and fetal, some hopeless, while others were helpless. Prognoses were generally poor. Threat compounded by threat.

That was 1983. I also recall being summoned to admit a patient by my supervising resident. He mentioned they were suspicious the patient had human T-cell lymphotropic virus (HTLV), something that had not been seen at our hospital before. [This virus would later be renamed Human Immunodeficiency Virus (HIV).] The patient was quite ill with pneumocystis carinii pneumonia and had an abnormal white blood cell count, suggesting immune dysfunction. The syndrome associated with this virus of wasting, opportunistic infections, and unusual cancers eventually leading to death, would later be named acquired immunodeficiency syndrome (AIDS).

I was only somewhat aware of the significance of this disease as well as the unknowns at that time. The virulence and transmissibility of this virus was still an unknown. But I was aware enough to be afraid. I repressed my fear. Stone faced and robotic, I proceeded to obtain a history, conduct an examination, draw multiple vials of blood, perform a lumbar puncture to obtain a sample of cerebrospinal fluid, write orders, and admit the patient to the hospital.

This was my introduction to clinical medicine—chronic incurable disease, disability, pain, suffering, and death. But let's not leave out my fear, fatigue, disrespect, repression, stoicism, and eventually, numbing with faltering to failing empathy and compassion. An insipid state of being. Fortunately, things got better for me, and they are generally better today for medical students than in 1983. Nevertheless, clinician burnout is approaching 50%, and suicides are at an all-time high within the profession. We still don't have this right, and it is affecting the care we provide.

Threat changes phenotype. Empathy and compassion are lost. [465] Things like teaching the concepts of empathy, pretending to be a patient to better understand the experience, and role-playing empathic behaviors are counterproductive. Role modeling empathy will not increase actual empathy. Empathy is not a cognitive construct to be acquired, learned, or measured, but an emotion to be felt, expressed, and transmitted. Safety restores empathy and compassion, along with kindness. Safety cures burnout.

I realize my career in medicine has the bookends of two horrific viruses, HIV and Covid-19. Both viruses presented the human species with significant threats and crises. Upon the discovery of both viruses, palpable fear and threat-based behaviors accompanied all the unknowns. HIV has killed approximately 40 million people world-wide over the past forty years. Covid-19 has killed at over 7 million people world-wide over the past four years. Neither virus has been eradicated. Both are now endemic in the human population. The panic resolved, the crises diffused, but the threat remains with us. What have we learned to help us with the next virus?

Both viruses disproportionately affected the disenfranchised. Different bugs at different times, yet not unique in sequelae. Covid-19 is neither the first nor the last pandemic.

The Bubonic Plague, a bacterial infection, killed approximately 200 million people in the 14th century. Disease and death in this pandemic disproportionately affected the poor and disenfranchised, especially those living in crowded unsanitary conditions.

The Spanish flu, a viral infection of the early 20th century, killed approximately 25-50 million people worldwide. A major pattern of mortality was the differences between social classes. These disparities reflected worse diets, poverty, crowded living conditions, and problems accessing healthcare in those with worse outcomes. Seems familiar.

This consistent pattern of the poor and disenfranchised suffering more and having worse outcomes, with pathogenic threats, repeats itself across the centuries and millennia.

Within the silos of medicine and science, we focus on these pathogens as the threats, but is there something else to see? Could these pathogens be unheard and unheralded messengers in addition to pathogens? Have their messages been delivered, yet gone unheeded?

I think, clearly these pathogens themselves have ignored human contracts, ideations, constructs, narratives, and beliefs. In fact, they are oblivious to human contracts, ideations, constructs, narratives, and beliefs. These pathogens see and use the likeness and connectivity of humans and the planet for surviving, thriving, and propagating their own genetic code. They disregard human color, race, religion, creed, gender, sexuality, nationality, and borders. They pick out those who are suffering the most and are already weakened from chronic threat in their viral infliction of damage and death. These pathogens point out the reality of the human condition, not the human contracts, ideations, constructs, narratives, and beliefs of the human condition. Biology is objective reality. We are all connected, and we are all the same, except for our levels of threat versus safety. Our differences are just subjective contracts, ideations, constructs, narratives, and beliefs of the human mind.

These pathogens have illuminated the other determinants of illness and disease versus wellness and health. People with higher TLs fare worse with these infections. Vaccines will help, but we should be targeting the other determinants of health, too. We should be targeting all the other threats. Isolating, masking, gowning, sanitizing, and chasing pathogens are not enough, and pose their own risks and harms. These strategies, as in the last pandemic, will not be enough in the next one.

In addition, these pathogens point out the phenotypes and physiologies of threat, both in the early or acute phase, as well as in the late or chronic phase. Acute threat can kill quickly, as seen with Covid-19 and influenza infections. The fight, flight, and fever of an acute infection unabated can devolve into the falter, faint, shock physiology, and cell, tissue, and organ damage, then possibly death, thus completing a destructive defense response across the entire threat spectrum—inflammation, oxidation, catabolism, immuno-paralysis, immobilization, cell, tissue, and organ injury, and in the end, death.

The acute Covid-19 process plays out in days to weeks. HIV is less aggressive in its onset, and prior to antivirals, death came, but more slowly than with Covid-19, yet the process was similar in sequence. Over months to years inflammation, oxidation, catabolism, immuno-paralysis, immobilization, cell, tissue, and organ injury, even cancer, all took their toll, and death followed.

Post Covid-19 syndrome, or Long Covid, may play out similarly, but in even slower motion and in the absence of the virus. In many of the chronic diseases and *syndromes* in medicine, this pattern of threat repeats and plays out over years and decades. The pattern even plays out in the rate of aging.

Can we look beyond the threat of these pathogens to see the sequelae are not just the pathogens alone? The pathogens simply speed things up to illuminate the processes more clearly within a shorter time frame. The sequelae are a result of the total TL and chronic threat accumulated and compounded from multiple sources eventually leading to chronic diseases, disabilities, pain and suffering, premature aging, and death—inflammation, oxidative stress, catabolism, immuno-paralysis, immobilization, cell, tissue, and organ injury, and cancer, followed by death.

It is the same root and the same soup of threat in different concentrations, doses, and virulences over different periods of time. We are experiencing the same process over and over. Yet, we are not seeing and understanding it completely, let alone treating it properly and completely.

One of my professors once told me, "Information does not change behavior." All the intellect we have and knowledge we acquire does not equate to wisdom. Wisdom is contingent on execution, not education. We now know, but are we wise?

"Bumble bees are greater experts on life than humans."

Ali F. Oeus

Studying these processes, the processes of a chronic unrelenting systemic TR, is essential to understand most chronic illness and disease. It is the entire exposome-genome interface that determines illness and disease versus wellness and health. Pathogens and biology can teach us the truth of our world and our condition. But we must step back, look up, and see beyond the single gene, toxin, mutagen, allergen, antigen, pathogen, and injury to perceive the entire picture.

Current research is heavily biased towards double blinded randomized controlled studies, usually targeting a single gene, toxin, mutagen, allergen, antigen, pathogen, and/or structure with a single intervention. These studies are designed to introduce an intervention or variable to be examined, while randomizing the subjects into the different groups. The hope is that all the other variables and determinants of outcome and health will wash out within the randomization of large groups. Thus, only the single variable, the intervention, requires analysis. Consider this type of analysis a small data construct to determine the efficacy of a specific intervention.

Historically, big data with a multiplicity of variables was too unwieldly to manage within scientific works. TVST contends that most answers to illness and disease versus wellness and health are not to be found in a single gene, toxin, mutagen, allergen, antigen, pathogen, or structure. Therefore, small data analysis should not be the priority in research nor clinical care. Technology has now evolved to be able to manage big data. Weaponized artificial intelligence can be a risk and a threat, but AI can also be used for good. AI allows us to design studies, not to wash out variables, but to include variables. All the variables of threat versus safety—generational, historical, ecological, physical, emotional, behavioral, social, mental, cultural, societal, institutional, financial, technological, spiritual—can be studied simultaneously for a better understanding of illness and disease versus wellness and health. Big data can thus be used to drive health care decisions and to prioritize health care funding and research. We can finally address 100% of the determinants of health—if we choose to.

We need a new paradigm, a global paradigm, for wellness and health. It must look beyond corporate profits, use our resources wisely and efficiently, and systematically decrease the TL of the world. At the same time, the paradigm must increase the SL throughout the world. We need clean air, pure water, and healthy food. We need access to healthcare, safe housing, and secure income. We need expressive freedom and safety in being able to express ourselves. We need close social connections, positive cognitive constructs, supportive cultural norms, reintegration with the natural world, and a restoring faith to displace rigid belief. We need to create the space and the time to heal and become resilient. We need all these things for everyone who walks on this planet. We are all suffering to different degrees from a Safety Deficit Disorder. We all need to feel and be safe.

"I would argue that a social justice approach should be central to medicine and utilized to be central to public health. This could be very simple: the well should take care of the sick."

Paul Farmer

There are tipping points of concern. One previously mention was when 2-3% of our cells become senescent cells they produce enough TS and TC that our immune cells and metabolic processes no longer seek out these senescent cells to deconstruct or destroy them. At this point, unabated, a precipitous decline in health can occur, followed by death.

Another tipping point we are approaching is in carbon emissions and global warming from which there may no longer be the ability to restore the planet to conditions necessary for sustaining human life. Humans run the risks of trillions of dollars in climate-related damage. Billions of people are being pushed into hardship around the world. Millions of lives are being lost because of a rapidly warming planet. Continued global warming will result in the loss of coral reefs, collapse of ice sheets, and catastrophic loss of crop-growing capacity. Thirty million children suffer from extreme weather-associated hunger today, and premature births and their complications are associated with the warming climate.

The changing climate is a major threat to humanity that will further accelerate disease, disability, pain, and suffering to compound an already excessive TL. Climate change is associated with worsening of sleep, dehydration, strokes, migraines, meningitis, epilepsy, multiple sclerosis, schizophrenia, Alzheimer's disease, and Parkinson's disease. The many components of climate change—air pollution, carbon gases, acidic waters, temperature rise, storms… promote threat physiology and the associated disease states.

As threat compounds if left unabated, I fear the global TL can also hit a tipping point where reversing threat phenotype, physiology, and behaviors may no longer be possible. Incivility, lawlessness, illness, disease, disability, pain, suffering, and deaths will follow.

Scientists suggest the mythical doomsday clock is now at 90 minutes to midnight—time is not reversible for humans. Our fate isn't sealed just yet, but action must come now to avert a mass hellscape or graveyard on Earth.

"Optimism isn't a belief that things will automatically get better; it's a conviction that we can make things better."

Melinda French Gates

We all need safety. We all need a safe home. Fundamentally Earth is our home. It has been shown that Earth systems, due to the impact of human activities, behave in a reactive and chaotic fashion.[466] Reciprocally, climate change chaos can influence the physical and mental health of humans.[467] It should not be assumed that the planet or the planet's human population are homeostatic systems that will seek an equilibrium. Certainly, the planet has a sweet spot from which biologic functions and life arise and at times thrive. But there is no inherent driver to this sweet spot. In fact, inorganicity appears to be the more common option for planets within our cosmos. Humans also have phenotypic sweet spots, but there isn't an inherent driver within humans to this spot, either. The exposome interfacing with the genome determines the phenotypic spot we land on—sweet or not.

Earth has her phenotypes, too, that play out both acutely and chronically. She has gone through stages and states of vast biological production and evolution, followed by multiple mass extinctions, dormancy, and infertility. Phases of extreme heat and extreme cold mark these biologically challenging periods. Some of the threats to life on Earth have been inorganic events, such as bombardments from outer space and volcanic eruptions. However, the organic has played a role as well. Approximately 2 billion years ago, toxic oxygen gases accounted for the extinction of 99.9% of the species on the planet and resulted in a profound cooling of the planet that lasted a billion years. This was an organic event, brought on by primitive cells evolving photosynthetic functions to use carbon dioxide and sunlight for fuel. Their waste production, oxygen gas, was overproduced and highly toxic to most lifeforms inhabiting Earth at the time. (Oxygen gases cause cooling, unlike the human waste products of carbon gases which cause warming.) For nearly a billion years following this photosynthetic bloom, Earth was toxic and frozen, and biologically relatively quiet to nearly dormant.

The impact of humans on Earth's systems cannot be understated. The human-driven production of carbon gases is destabilizing and is pushing the environment towards a chaotic hothouse, from which there may be no return for millions or perhaps billions of years. How many species will survive the extreme heat and extreme weather that is to come? Without intervention, humans will certainly perish. Within this massive threat now upon us, will there be massive dissolution, with only primitive single-cell organisms surviving this extreme?

Earth is on the threat spectrum—fight, flight, fever, freeze, falter—will faint follow?

Cumulative and compounded ecological threats are truly existential threats for humans, but reciprocally humans are a threat to the stability of the planet. Although humans are a major threat to Earth, Earth has powerful defenses, immunologic and metabolic programs, to thwart this threat. From anger through despair, boundaries will be set for humans and defenses deployed against humans.

Earth now appears to be in a T1 phenotype. Angry windstorms cause mass destruction. Intense and explosive fires scorch and scar the land killing everything in their paths. At times, Earth weeps in a deluge of rains and floods, cleansing a soul in hopes of a better tomorrow. Although it is frightening to humans, Earth within its immunologic and metabolic strategies displays its wisdom within the windstorms, faith within the forest fires, and attempts at resurrection within the flooding rains. However, if this human sepsis isn't controlled by these methods Earth may retreat into a T2 phenotype to become listless and lifeless again. Will this state be the ultimate cure for the human pandemic, Earth's current plague?

Earth will survive and thrive again, but Earth may choose to move onward without humans. Alternately, humans could adapt, cooperate, collaborate, and survive, hopefully thrive, as a symbiote on planet Earth.

Can we not only make each other feel safe, but make our home safe, Make Earth Safe Again?

This is our next moonshot—no small feat, but absolutely doable—only this time it is an Earth-shot that can't afford to fail.

Do we really want to continue in an old and dysfunctional paradigm?

Insanity:
Doing the same thing over and over again
and expecting different results

Albert Einstein

Big Bang Principles of TVST

Life is energy, motion, and creation that needs boundaries, defenses, deconstruction, reconstruction, reproduction, recovery, nutrition, connection, participation...and safety.

1. Good in (nutrition, connections) – Bad out (pollutants, toxins)
2. Breath and move (physically, emotionally, mentally)
3. Stretch and be flexible (physically, emotionally, mentally)
4. Express and release emotions and thoughts
5. Embrace music and dance
6. Participate in play and labor
7. Nourish curiosity and creativity
8. Resist automatic narratives and rigid beliefs
9. Have boundaries and respect boundaries
10. Be accountable and hold to account
11. Connect with safe others (people, animals, nature)
12. Disconnect from technologies (screens, media, marketing)
13. Experience the freedom of wandering in mind and body
14. Engage in deconstruction and reconstruction
15. Value rest and recovery
16. Sleep well and enough
17. Lead with service and sacrifice
18. Build a culture founded on safety
19. Support safety in institutions, policies, procedures, laws, people, and actions
20. Vote for policies, procedures, laws, and people that are committed to increasing safety in the world

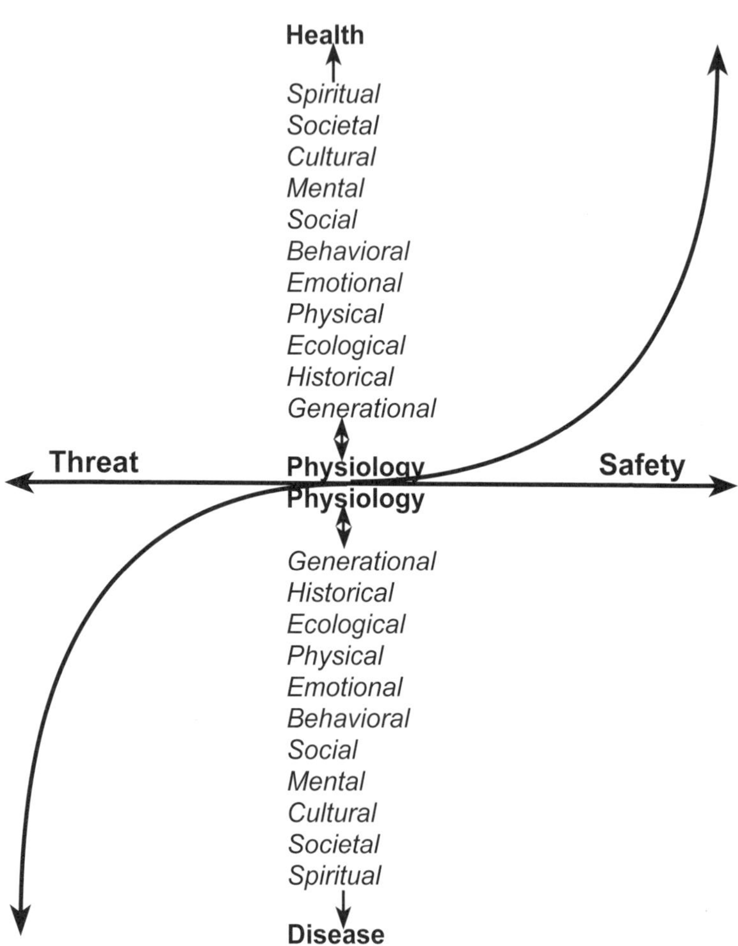
Health
Spiritual
Societal
Cultural
Mental
Social
Behavioral
Emotional
Physical
Ecological
Historical
Generational
Threat
Physiology
Safety
Physiology
Generational
Historical
Ecological
Physical
Emotional
Behavioral
Social
Mental
Cultural
Societal
Spiritual
Disease

"Medical statistics will be our standard of measurement: we will weigh life for life and see where the dead lie thicker, among the workers or among the privileged."

Rudolf Virchow

References

1. Chassin MR, Loeb JM. High-reliability health care: getting there from here. The Milbank Quarterly. 2013 Sep;91(3):459-90.

2. Bongaarts, L. National Research Council, Committee on Population. U.S. health in international perspective: Shorter lives, poorer health. 2013.

3. Amadeo, K. Medical Bankruptcy and the Economy – Do Medical Bills Really Devastate Families. The Balance. 2022, Jan 22.

4. Heuveline P. The Covid-19 pandemic and the expansion of the mortality gap between the United States and its European peers. PLoS One. 2023 Mar 29;18(3):e0283153.

5. Gunja MZ, Gumas ED, Williams RD. U.S. health care from a global perspective, 2022: accelerating spending, worsening outcomes. The Commonwealth Fund. Published January. 2023;21.

6. Dworak EM, Revelle W, Condon DM. Looking for Flynn effects in a recent online U.S. adult sample: Examining shifts within the SAPA Project. Intelligence. 2023 May 1;98:101734.

7. Zimorski V, Ku C, Martin WF, Gould SB. Endosymbiotic theory for organelle origins. Current opinion in microbiology. 2014 Dec 1;22:38-48.

8. Deamer D. The role of lipid membranes in life's origin. Life. 2017 Jan 17;7(1):5.

9. McNichol J. Primordial soup, fool's gold, and spontaneous generation: A brief introduction to the theory, history, and philosophy of the

search for the origin of life. Biochemistry and Molecular Biology Education. 2008 Jul;36(4):255-61.

10. Nielsen PE. Peptide nucleic acids and the origin of life. Chemistry & Biodiversity. 2007 Sep;4(9):1996-2002.

11. Makarov M, Sanchez Rocha AC, Krystufek R, Cherepashuk I, Dzmitruk V, Charnavets T, Faustino AM, Lebl M, Fujishima K, Fried SD, Hlouchova K. Early selection of the amino acid alphabet was adaptively shaped by biophysical constraints of foldability. Journal of the American Chemical Society. 2023 Feb 24;145(9):5320-9.

12. Wimmer JL, Xavier JC, Vieira AD, Pereira DP, Leidner J, Sousa FL, Kleinermanns K, Preiner M, Martin WF. Energy at origins: favorable thermodynamics of biosynthetic reactions in the last universal common ancestor (LUCA). Frontiers in microbiology. 2021:3903.

13. Chatterjee S, Yadav S. The origin of prebiotic information system in the peptide/RNA world: a simulation model of the evolution of translation and the genetic code. Life. 2019 Mar 1;9(1):25.

14. LeRoux M, Peterson SB, Mougous JD. Bacterial danger sensing. Journal of Molecular Biology. 2015 Nov 20;427(23):3744-53.

15. Kausar S, Yang L, Abbas MN, Hu X, Zhao Y, Zhu Y, Cui H. Mitochondrial DNA: a key regulator of anti-microbial innate immunity. Genes. 2020 Jan 11;11(1):86.

16. Naviaux RK. Metabolic features of the cell danger response. Mitochondrion. 2014 May 1;16:7-17.

17. Verbeke F, De Craemer S, Debunne N, Janssens Y, Wynendaele E, Van de Wiele C, De Spiegeleer B. Peptides as quorum sensing molecules: measurement techniques and obtained levels in vitro and in vivo. Frontiers in Neuroscience. 2017 Apr 12;11:183.

18. Antonioli L, Blandizzi C, Pacher P, Guilliams M, Haskó G. Rethinking communication in the immune system: the quorum sensing concept. Trends in Immunology. 2019 Feb 1;40(2):88-97.

19. Moffett AS, Thomas PJ, Hinczewski M, Eckford AW. Cheater suppression and stochastic clearance through quorum sensing. PLOS Computational Biology. 2022 Jul 28;18(7):e1010292.

20. Yutin N, Wolf MY, Wolf YI, Koonin EV. The origins of phagocytosis and eukaryogenesis. Biology Direct. 2009 Dec;4(1):1-26.

21. Morono Y, Ito M, Hoshino T, Terada T, Hori T, Ikehara M, D'Hondt S, Inagaki F. Aerobic microbial life persists in oxic marine sediment as old as 101.5 million years. Nature Communications. 2020 Jul 28;11(1):1-9.

22. Litman GW, Cannon JP, Dishaw LJ. Reconstructing immune phylogeny: new perspectives. Nature Reviews Immunology. 2005 Nov;5(11):866-79.

23. Koonin EV, Makarova KS, Wolf YI. Evolutionary genomics of defense systems in archaea and bacteria. Annual Review of Microbiology. 2017 Sep 8;71:233.

24. Calfee CS. Analysis of immunoparalysis. Science Translational Medicine. 2012 Nov 14;4(160):160ec208-.

25. Ameisen JC. On the origin, evolution, and nature of programmed cell death: a timeline of four billion years. Cell Death & Differentiation. 2002 Apr;9(4):367-93.

26. Dr. Taryn Marie Stejskal Shares the New Rules of Resilience with Division of Programs in Business. NYU School of Professional Studies. 2020 Nov 10.

27. Howick J, Dudko M, Feng SN, Ahmed AA, Alluri N, Nockels K, Winter R, Holland R. Why might medical student empathy change throughout medical school? a systematic review and thematic synthesis of qualitative studies. BMC Medical Education. 2023 Dec;23(1):1-3.

28. Neumann M, Edelhäuser F, Tauschel D, Fischer MR, Wirtz M, Woopen C, Haramati A, Scheffer C. Empathy decline and its reasons: a systematic review of studies with medical students and residents. Academic Medicine. 2011 Aug 1;86(8):996-1009.

29. Billman GE. Homeostasis: the underappreciated and far too often ignored central organizing principle of physiology. Frontiers in Physiology. 2020 Mar 10;11:200.

30. McEwen BS, Stellar E. Stress and the individual: Mechanisms leading to disease. Archives of Internal Medicine. 1993 Sep 27;153(18):2093-101.

31. McEwen BS. Stressed or stressed out: what is the difference? Journal of Psychiatry and Neuroscience. 2005 Sep 1;30(5):315-8.

32. Jean-Philippe Gouin, C. Sue Carter, et al Marital Behavior, Oxytocin, Vasopressin, and Wound Healing. Psychoneuroendocrinology. 2010 Aug; 35(7): 1082–1090.

33. Crawford S, Schold J. Association between geographic measures of socioeconomic status and deprivation and major surgical outcomes. Medical Care. 2019 Dec 15;57(12):949-59.

34. Jabr F. How brainless slime molds redefine intelligence. Nat. News. (2012).

35. Botton-Amiot G, Martinez P, Sprecher SG. Associative learning in the cnidarian Nematostella vectensis. Proceedings of the National Academy of Sciences. 2023 Mar 28;120(13):e2220685120.

36. Godfrey-Smith P. Other minds: The octopus, the sea, and the deep origins of consciousness. Farrar, Straus and Giroux; 2016 Dec 6.

37. Koonin EV, Wolf YI. Constraints and plasticity in genome and molecular-phenome evolution. Nature Reviews Genetics. 2010 Jul;11(7):487-98.

38. Davey CG, Harrison BJ. The self on its axis: a framework for understanding depression. Translational Psychiatry. 2022 Jan 18;12(1):1-9.

39. Solomon S. The role of death denial in culture and consciousness. Emerging Trends in the Social and Behavioral Sciences: An Interdisciplinary, Searchable, and Linkable Resource. 2015 May 15:1-6.

40. Solomon S, Greenberg J, Pyszczynski T. Pride and prejudice: Fear of death and social behavior. Current Directions in Psychological Science. 2000 Dec;9(6):200-4.

41. Leary MR. The curse of the self: Self-awareness, egotism, and the quality of human life. Oxford University Press; 2007.

42. Seth AK, Bayne T. Theories of consciousness. Nature Reviews Neuroscience. 2022 May 3:1-4.

43. Thomaes S, Brummelman E, Reijntjes A, Bushman BJ. When Narcissus was a boy: Origins, nature, and consequences of childhood narcissism. Child Development Perspectives. 2013 Mar;7(1):22-6.

44. Amster, M and Eagle, J. Amster, Stuck at Home? How to Find Awe and Beauty Indoors. APRIL 15, 2020.

45. Nutt D, Carhart-Harris R. The current status of psychedelics in psychiatry. JAMA psychiatry. 2021 Feb 1;78(2):121-2.

46. Moliner R, Girych M, Brunello CA, Kovaleva V, Biojone C, Enkavi G, Antenucci L, Kot EF, Goncharuk SA, Kaurinkoski K, Kuutti M. Psychedelics promote plasticity by directly binding to BDNF receptor TrkB. Nature Neuroscience. 2023 Jun;26(6):1032-41.

47. Nannini DR, Zheng Y, Joyce BT, Kim K, Gao T, Wang J, Jacobs DR, Schreiner PJ, Yaffe K, Greenland P, Lloyd-Jones DM. Genome-wide DNA methylation association study of recent and cumulative marijuana use in middle aged adults. Molecular Psychiatry. 2023 May 31:1-1.

48. Calder AE, Hasler G. Towards an understanding of psychedelic-induced neuroplasticity. Neuropsychopharmacology. 2023 Jan;48(1):104-12.

49. Monroy M, Keltner D. Awe as a pathway to mental and physical health. Perspectives on Psychological Science. 2023 Mar;18(2):309-20.

50. Kerskens CM, Perez DL. Experimental indications of non-classical brain functions. Journal of Physics Communications. 2022 Sep 26.

51. DeRijk RH, van Leeuwen N, Klok MD, Zitman FG. Corticosteroid receptor-gene variants: modulators of the stress-response and implications for mental health. European Journal of Pharmacology. 2008 May 13;585(2-3):492-501.

52. Yin H, Pickering JG. Telomere Length: Implications for Atherogenesis. Current Atherosclerosis Reports. 2023 Jan 23:1-9.

53. Felitti V, Anda R, Nordenberg D, et al. Relationship of childhood abuse and household dysfunction to many of the leading causes of death in adults. The Adverse Childhood Experiences (ACE) Study. Am J Prev Med 1998 May;14(4):245-58.

54. Brudey C, Park J, Wiaderkiewicz J, Kobayashi I, Mellman TA, Marvar PJ. Autonomic and inflammatory consequences of posttraumatic stress disorder and the link to cardiovascular disease. American Journal of Physiology-Regulatory, Integrative and Comparative Physiology. 2015 Aug 15;309(4):R315-21.

55. Youssef, N. A., Lockwood, L., Su, S., Hao, G., & Rutten, B. (2018). The Effects of Trauma, with or without PTSD, on the Transgenerational DNA Methylation Alterations in Human Offsprings. *Brain Sciences*, *8*(5), 83.

56. Yin W, Ludvigsson JF, Åden U, Risnes K, Persson M, Reichenberg A, Silverman ME, Kajantie E, Sandin S. Paternal and maternal psychiatric history and risk of preterm and early term birth: A nationwide study using Swedish registers. PLoS medicine. 2023 Jul 20;20(7):e1004256.

57. Barrett LF, Simmons WK. Interoceptive predictions in the brain. Nature reviews neuroscience. 2015 Jul;16(7):419-29.

58. Naviaux RK. Perspective: Cell danger response Biology—The new science that connects environmental health with mitochondria and the rising tide of chronic illness. Mitochondrion. 2020 Mar 1;51:40-5.

59. Otten M, Seth AK, Pinto Y. Seeing Ɔ, remembering C: Illusions in short-term memory. Plos One. 2023 Apr 5;18(4):e0283257.

60. Berlin HA. The neural basis of the dynamic unconscious. Neuropsychoanalysis. 2011 Jan 1;13(1):5-31.

61. Revonsuo A. The reinterpretation of dreams: An evolutionary hypothesis of the function of dreaming. Behavioral and Brain Sciences. 2000 Dec;23(6):877-901.Berlin HA. The neural basis of the dynamic unconscious. Neuropsychoanalysis. 2011 Jan 1;13(1):5-31.

62. Francis L. Stevens (2016) The anterior cingulate cortex in psychopathology and psychotherapy: effects on awareness and repression of affect, Neuropsychoanalysis, 18:1, 53-68.

63. Sarno J. The Divided Mind: The Epidemic of Mindbody Disorders. (2006). Harper ISBN 0-06-085178-3.

64. Stevens JS, Harnett NG, Lebois LA, Van Rooij SJ, Ely TD, Roeckner A, Vincent N, Beaudoin FL, An X, Zeng D, Neylan TC. Brain-based biotypes of psychiatric vulnerability in the acute aftermath of trauma. American Journal of Psychiatry. 2021 Nov;178(11):1037-49.

65. McEwen CA. Connecting the biology of stress, allostatic load and epigenetics to social structures and processes. Neurobiology of Stress. 2022 Mar 1;17:100426.

66. Bianchi ME. DAMPs, PAMPs and alarmins: all we need to know about danger. Journal of Leukocyte Biology. 2007 Jan;81(1):1-5.

67. Yazici D, Ogulur I, Kucukkase O, Li M, Rinaldi AO, Pat Y, Wallimann A, Wawrocki S, Celebi Sozener Z, Buyuktiryaki B, Sackesen C. Epi-

thelial barrier hypothesis and the development of allergic and autoimmune diseases. Allergo Journal International. 2022 May 4:1-2.

68. Akdis CA. Does the epithelial barrier hypothesis explain the increase in allergy, autoimmunity and other chronic conditions? Nature Reviews Immunology. 2021 Nov;21(11):739-51.

69. Pat Y, Ogulur I. The epithelial barrier hypothesis: a 20-year journey. Allergy. 2021 Nov 1;76(11):3560-2.

70. U.S. Cancer Statistics Working Group. United States cancer statistics: 1999–2014 incidence and mortality web-based report. Atlanta: U.S. Department of Health and Human Services, Centers for Disease Control and Prevention and National Cancer Institute. 2017 Aug 9;201.

71. Jackson JH. The Croonian lectures on evolution and dissolution of the nervous system. British Medical Journal. 1884 Apr 4;1(1215):703.

72. Mercier C. Hughlings-Jackson on Evolution and Dissolution of the Nervous System. Brain. 1884 Jul 1;7(2):283-4.

73. Fan Y, Ullman E, Zong WX. The cellular decision between apoptosis and autophagy. Beyond Apoptosis. 2008 Sep 19:141-56.

74. Alexandra Kredlow M, Fenster RJ, Laurent ES, Ressler KJ, Phelps EA. Prefrontal cortex, amygdala, and threat processing: implications for PTSD. Neuropsychopharmacology. 2022 Jan;47(1):247-59.

75. Jeong H, Chung YA, Ma J, Kim J, Hong G, Oh JK, Kim M, Ha E, Hong H, Yoon S, Lyoo IK. Diverging roles of the anterior insula in trauma-exposed individuals vulnerable or resilient to posttraumatic stress disorder. Scientific Reports. 2019 Oct 29;9(1):1-8.

76. Brennan KC, Pietrobon D. A systems neuroscience approach to migraine. Neuron. 2018 Mar 7;97(5):1004-21.

77. Rafiq S, Campodonico C, Varese F. The relationship between childhood adversities and dissociation in severe mental illness: A meta-analytic review. Acta Psychiatrica Scandinavica. 2018 Dec;138(6):509-25.

78. Hofford RS, Russo SJ, Kiraly DD. Neuroimmune mechanisms of psychostimulant and opioid use disorders. European Journal of Neuroscience. 2019 Aug;50(3):2562-73.

79. Allen J, Romay-Tallon R, Brymer KJ, Caruncho HJ, Kalynchuk LE. Mitochondria and mood: mitochondrial dysfunction as a key player in the manifestation of depression. Frontiers in Neuroscience. 2018 Jun 6;12:386.

80. Koss MP, Marks JS. Relationship of childhood abuse and household dysfunction to many of the leading causes of death in adults: The adverse childhood experiences (ACE) study. American Journal of Preventative Medicine. 1998 May;14(4):245-58.

81. Dai W, Jiang L. Dysregulated mitochondrial dynamics and metabolism in obesity, diabetes, and cancer. Frontiers in Endocrinology. 2019 Sep 3;10:570.

82. Ekene OC. Cytokines and HIV/AIDS: a critical look at the existing relationship between them. Roumanian Archives of Microbiology and Immunology. 2008 Jul 1;67(3-4):67-80.

83. Bhat AA, Nisar S, Maacha S, Carneiro-Lobo TC, Akhtar S, Siveen KS, Wani NA, Rizwan A, Bagga P, Singh M, Reddy R. Cytokine-chemokine network driven metastasis in esophageal cancer; promising avenue for targeted therapy. Molecular Cancer. 2021 Dec;20(1):1-20.

84. Lin YT, Seo J, Gao F, Feldman HM, Wen HL, Penney J, Cam HP, Gjoneska E, Raja WK, Cheng J, Rueda R. APOE4 causes widespread molecular and cellular alterations associated with Alzheimer's disease phenotypes in human iPSC-derived brain cell types. Neuron. 2018 Jun 27;98(6):1141-54.

85. Garcia AR, Finch C, Gatz M, Kraft T, Rodriguez DE, Cummings D, Charifson M, Buetow K, Beheim BA, Allayee H, Thomas GS. APOE4 is associated with elevated blood lipids and lower levels of innate immune biomarkers in a tropical Amerindian subsistence population. Elife. 2021 Sep 29;10:e68231.

86. Laverty G, Gorman SP, Gilmore BF. Biofilms and implant-associated infections. Biomaterial and Medical Device Associated Infection. 2015:19-45.

87. Krasnokutski SA, Chuang KJ, Jäger C, Ueberschaar N, Henning T. A pathway to peptides in space through the condensation of atomic carbon. Nature Astronomy. 2022 Mar;6(3):381-6.

88. Chatterjee S, Yadav S. The origin of prebiotic information system in the peptide/RNA world: a simulation model of the evolution of translation and the genetic code. Life. 2019 Mar 1;9(1):25.

89. Burton AS, Stern JC, Elsila JE, Glavin DP, Dworkin JP. Understanding prebiotic chemistry through the analysis of extraterrestrial amino acids and nucleobases in meteorites. Chemical Society Reviews. 2012;41(16):5459-72.

90. Naviaux RK. Metabolic features of the cell danger response. Mitochondrion. 2014 May 1;16:7-17.

91. Carter CS, Kenkel WM, MacLean EL, Wilson SR, Perkeybile AM, Yee JR, Ferris CF, Nazarloo HP, Porges SW, Davis JM, Connelly JJ. Is oxytocin "nature's medicine"?. Pharmacological Reviews. 2020 Oct 1;72(4):829-61.

92. Joëls M, Baram TZ. The neuro-symphony of stress. Nature Reviews Neuroscience. 2009 Jun;10(6):459-66.

93. Slavich GM, Sacher J. Stress, sex hormones, inflammation, and major depressive disorder: Extending Social Signal Transduction Theory of Depression to account for sex differences in mood disorders. Psychopharmacology. 2019 Oct;236(10):3063-79.

94. Dinarello CA. Historical insights into cytokines. European Journal of Immunology. 2007 Nov;37(S1):S34-45.

95. Ramani T, Auletta CS, Weinstock D, Mounho-Zamora B, Ryan PC, Salcedo TW, Bannish G. Cytokines: the good, the bad, and the deadly. International Journal of Toxicology. 2015 Jul;34(4):355-65.

96. Jaffer U, Wade RG, Gourlay T. Cytokines in the systemic inflammatory response syndrome: a review. HSR Proceedings in Intensive Care & Cardiovascular Anesthesia. 2010;2(3):161.

97. Gulati K, Guhathakurta S, Joshi J, Rai N, Ray AJ. Cytokines and their role in health and disease: a brief overview. Moj Immunol. 2016;4(2):1-9.

98. Giuliani C, Bucci I, Napolitano G. The role of the transcription factor nuclear factor-kappa B in thyroid autoimmunity and cancer. Frontiers in Endocrinology. 2018 Aug 21;9:471.

99. Hosokawa H, Rothenberg EV. Cytokines, transcription factors, and the initiation of T-cell development. Cold Spring Harbor Perspectives in Biology. 2018 May 1;10(5):a028621.

100. Capaldo CT, Nusrat A. Cytokine regulation of tight junctions. Biochimica et Biophysica Acta (BBA)-Biomembranes. 2009 Apr 1;1788(4):864-71.

101. Savla SR, Prabhavalkar KS, Bhatt LK. Cytokine storm associated coagulation complications in COVID-19 patients: Pathogenesis and Management. Expert review of anti-infective therapy. 2021 Nov 2;19(11):1397-413.

102. Kastl L, Sauer SW, Ruppert T, Beissbarth T, Becker MS, Süss D, Krammer PH, Gülow K. TNF-a mediates mitochondrial uncoupling and enhances ROS-dependent cell migration via NF-kB activation in liver cells. FEBS Letters. 2014 Jan 3;588(1):175-83.

103. Harris J. Autophagy and cytokines. Cytokine. 2011 Nov 1;56(2):140-4.

104. Sharma R, Anker SD. Cytokines, apoptosis and cachexia: the potential for TNF antagonism. International Journal of Cardiology. 2002 Sep 1;85(1):161-71.

105. Bertheloot D, Latz E, Franklin BS. Necroptosis, pyroptosis and apoptosis: an intricate game of cell death. Cellular & Molecular Immunology. 2021 May;18(5):1106-21.

106. Sookoian S, Pirola CJ. Alanine and aspartate aminotransferase and glutamine-cycling pathway: their roles in pathogenesis of metabolic syndrome. World Journal of Gastroenterology: WJG. 2012 Aug 8;18(29):3775.

107. Pribiag H, Stellwagen D. TNF-a downregulates inhibitory neurotransmission through protein phosphatase 1-dependent trafficking of GABAA receptors. Journal of Neuroscience. 2013 Oct 2;33(40):15879-93.

108. Clark IA, Vissel B. Excess cerebral TNF causing glutamate excitotoxicity rationalizes treatment of neurodegenerative diseases and neurogenic pain by anti-TNF agents. Journal of Neuroinflammation. 2016 Dec;13(1):1-6.

109. Ye L, Huang Y, Zhao L, Li Y, Sun L, Zhou Y, Qian G, Zheng JC. IL-1b and TNF-a induce neurotoxicity through glutamate production: a potential role for neuronal glutaminase. Journal of Neurochemistry. 2013 Jun;125(6):897-908.

110. Miller AH, Haroon E, Raison CL, Felger JC. Cytokine targets in the brain: impact on neurotransmitters and neurocircuits. Depression and Anxiety. 2013 Apr;30(4):297-306.

111. Vancassel S, Capuron L, Castanon N. Brain kynurenine and BH4 pathways: relevance to the pathophysiology and treatment of inflamma-

tion-driven depressive symptoms. Frontiers in Neuroscience. 2018 Jul 24;12:499.

112. Kenney MJ, Ganta CK. Autonomic nervous system and immune system interactions. Comprehensive physiology. 2014 Jul;4(3):1177.

113. Woo E, Sansing LH, Arnsten AF, Datta D. Chronic stress weakens connectivity in the prefrontal cortex: Architectural and molecular changes. Chronic Stress. 2021 Aug;5:24705470211029254.

114. Johnson JD, Barnard DF, Kulp AC, Mehta DM. Neuroendocrine regulation of brain cytokines after psychological stress. Journal of the Endocrine Society. 2019 Jul;3(7):1302-20.

115. Kariagina A, Romanenko D, Ren SG, Chesnokova V. Hypothalamic-pituitary cytokine network. Endocrinology. 2004 Jan 1;145(1):104-12.

116. Maydych V, Claus M, Watzl C, Kleinsorge T. Attention to emotional information is associated with cytokine responses to psychological stress. Frontiers in Neuroscience. 2018 Oct 2;12:687.

117. Maydych V. The interplay between stress, inflammation, and emotional attention: relevance for depression. Frontiers in Neuroscience. 2019 Apr 24;13:384.

118. Cooper CM, Godlewska B, Sharpley AL, Barnes E, Cowen PJ, Harmer CJ. Interferon-a induces negative biases in emotional processing in patients with hepatitis C virus infection: a preliminary study. Psychological Medicine. 2018 Apr;48(6):998-1007.

119. Slavich GM, Irwin MR. From stress to inflammation and major depressive disorder: a social signal transduction theory of depression. Psychological Bulletin. 2014 May;140(3):774.

120. Inagaki TK, Muscatell KA, Irwin MR, Cole SW, Eisenberger NI. Inflammation selectively enhances amygdala activity to socially threatening images. Neuroimage. 2012 Feb 15;59(4):3222-6.

121. Muscatell KA, Dedovic K, Slavich GM, Jarcho MR, Breen EC, Bower JE, Irwin MR, Eisenberger NI. Greater amygdala activity and dorsomedial prefrontal–amygdala coupling are associated with enhanced inflammatory responses to stress. Brain, Behavior, and Immunity. 2015 Jan 1;43:46-53.

122. Harrison NA, Brydon L, Walker C, Gray MA, Steptoe A, Critchley HD. Inflammation causes mood changes through alterations in subgenual cingulate activity and mesolimbic connectivity. Biological Psychiatry. 2009 Sep 1;66(5):407-14.

123. Gray MA, Chao CY, Staudacher HM, Kolosky NA, Talley NJ, Holtmann G. Anti-TNFa therapy in IBD alters brain activity reflecting visceral sensory function and cognitive-affective biases. PLoS One. 2018 Mar 8;13(3):e0193542.

124. Eshel N, Roiser JP. Reward and punishment processing in depression. Biological Psychiatry. 2010 Jul 15;68(2):118-24.

125. Harrison NA, Voon V, Cercignani M, Cooper EA, Pessiglione M, Critchley HD. A neurocomputational account of how inflammation enhances sensitivity to punishments versus rewards. Biological Psychiatry. 2016 Jul 1;80(1):73-81.

126. Shi J, Fan J, Su Q, Yang Z. Cytokines and abnormal glucose and lipid metabolism. Front Endocrinol (Lausanne) 10: 703.

127. Zoico E, Roubenoff R. The role of cytokines in regulating protein metabolism and muscle function. Nutrition Reviews. 2002 Feb 1;60(2):39-51.

128. Al-Mansoori L, Al-Jaber H, Prince MS, Elrayess MA. Role of inflammatory cytokines, growth factors and adipokines in adipogenesis and insulin resistance. Inflammation. 2021 Sep 18:1-4.

129. Feingold KR. Soued M, Adi S, Staprans I, Shigenaga J, Doerrler W, Moser A, Grunfeld C. Tumor necrosis factor-increased hepatic very-low-density lipoprotein production and increased serum triglyceride levels in diabetic rats. Diabetes. 1990;39:1569-74.

130. Nachiappan V, Curtiss D, Corkey BE, Kilpatrick L. Cytokines inhibit fatty acid oxidation in isolated rat hepatocytes: synergy among TNF, IL-6, and IL-1. Shock (Augusta, Ga.). 1994 Feb 1;1(2):123-9.

131. Yang D, Elner SG, Bian ZM, Till GO, Petty HR, Elner VM. Pro-inflammatory cytokines increase reactive oxygen species through mitochondria and NADPH oxidase in cultured RPE cells. Experimental Eye Research. 2007 Oct 1;85(4):462-72.

132. Seth A, Stemple DL, Barroso I. The emerging use of zebrafish to model metabolic disease. Disease Models & Mechanisms. 2013 Sep;6(5):1080-8.

133. Tracey KJ, Lowry SF, Cerami A. Cachectin: a hormone that triggers acute shock and chronic cachexia. The Journal of Infectious Diseases. 1988 Mar 1:413-20.

134. Kwilasz AJ, Grace PM, Serbedzija P, Maier SF, Watkins LR. The therapeutic potential of interleukin-10 in neuroimmune diseases. Neuropharmacology. 2015 Sep 1;96:55-69.

135. Ip WE, Hoshi N, Shouval DS, Snapper S, Medzhitov R. Anti-inflammatory effect of IL-10 mediated by metabolic reprogramming of macrophages. Science. 2017 May 5;356(6337):513-9.

136. Kobayashi T, Matsuoka K, Sheikh SZ, Russo SM, Mishima Y, Collins C, DeZoeten EF, Karp CL, Ting JP, Sartor RB, Plevy SE. IL-10 regulates Il12b expression via histone deacetylation: implications for intestinal macrophage homeostasis. The Journal of Immunology. 2012 Aug 15;189(4):1792-9.

137. Makita N, Hizukuri Y, Yamashiro K, Murakawa M, Hayashi Y. IL-10 enhances the phenotype of M2 macrophages induced by IL-4 and confers the ability to increase eosinophil migration. International Immunology. 2015 Mar 1;27(3):131-41.

138. Pullisaar H, Colaianni G, Lian AM, Vandevska-Radunovic V, Grano M, Reseland JE. Irisin promotes growth, migration and matrix formation in human periodontal ligament cells. Archives of Oral Biology. 2020 Mar 1;111:104635.

139. Wang Y, Tian M, Tan J, et al. Irisin ameliorates neuroinflammation and neuronal apoptosis through integrin aVb5/AMPK signaling pathway after intracerebral hemorrhage in mice. J Neuroinflammation. 2022;19(1):82.

140. Huang YS, Ogbechi J, Clanchy FI, Williams RO, Stone TW. IDO and kynurenine metabolites in peripheral and CNS disorders. Frontiers in Immunology. 2020 Mar 5;11:388.

141. Córdoba-Moreno MO, de Souza ED, Quiles CL, dos Santos-Silva D, Kinker GS, Muxel SM, Markus RP, Fernandes PA. Rhythmic expression of the melatonergic biosynthetic pathway and its differential modulation in vitro by LPS and IL10 in bone marrow and spleen. Scientific Reports. 2020 Mar 16;10(1):1-2.

142. Patel RR, Wolfe SA, Bajo M, Abeynaike S, Pahng A, Borgonetti V, D'Ambrosio S, Nikzad R, Edwards S, Paust S, Roberts AJ. IL-10 normalizes aberrant amygdala GABA transmission and reverses anxiety-like behavior and dependence-induced escalation of alcohol intake. Progress in Neurobiology. 2021 Apr 1;199:101952.

143. Pesce M, La Fratta I, Paolucci T, Grilli A, Patruno A, Agostini F, Bernetti A, Mangone M, Paoloni M, Invernizzi M, De Sire A. From exercise to cognitive performance: Role of irisin. Applied Sciences. 2021 Jul 31;11(15):7120.

144. Tekin S, Beytur A, Erden Y, Beytur A, Cigremis Y, Vardi N, Turkoz Y, Tekedereli I, Sandal S. Effects of intracerebroventricular administration of irisin on the hypothalamus–pituitary–gonadal axis in male rats. Journal of Cellular Physiology. 2019 Jun;234(6):8815-24.

145. Küster OC, Laptinskaya D, Fissler P, Schnack C, Zügel M, Nold V, Thurm F, Pleiner S, Karabatsiakis A, von Einem B, Weydt P. Novel blood-based biomarkers of cognition, stress, and physical or cognitive training in older adults at risk of dementia: preliminary evidence for a role of BDNF, irisin, and the kynurenine pathway. Journal of Alzheimer's Disease. 2017 Jan 1;59(3):1097-111.

146. Minton K. IL-10 targets macrophage metabolism. Nature Reviews Immunology. 2017 Jun;17(6):345-.

147. Priyadharshini B, Loschi M, Newton RH, Zhang JW, Finn KK, Gerriets VA, Huynh A, Rathmell JC, Blazar BR, Turka LA. Cutting edge: TGF-b and phosphatidylinositol 3-kinase signals modulate distinct metabolism of regulatory T cell subsets. The Journal of Immunology. 2018 Oct 15;201(8):2215-9.

148. Fu J, Li F, Tang Y, Cai L, Zeng C, Yang Y, Yang J. The Emerging Role of Irisin in Cardiovascular Diseases. Journal of the American Heart Association. 2021 Oct 19;10(20):e022453.

149. Luo Y, Qiao X, Ma Y, Deng H, Xu CC, Xu L. Disordered metabolism in mice lacking irisin. Scientific Reports. 2020 Oct 15;10(1):1-0.

150. Straczkowski M, Kowalska I, Nikolajuk A, Krukowska A, Gorska M. Plasma interleukin-10 concentration is positively related to insulin sensitivity in young healthy individuals. Diabetes Care. 2005 Aug 1;28(8):2036-7.

151. Caligiuri G, Rudling M, Ollivier V, Jacob MP, Michel JB, Hansson

GK, Nicoletti A. Interleukin-10 deficiency increases atherosclerosis, thrombosis, and low-density lipoproteins in apolipoprotein E knockout mice. Molecular Medicine. 2003 Jan;9(1):10-7.

152. Gómez-Sámano MÁ, Grajales-Gómez M, Zuarth-Vázquez JM, Navarro-Flores MF, Martínez-Saavedra M, Juárez-León ÓA, Morales-García MG, Enríquez-Estrada VM, Gómez-Pérez FJ, Cuevas-Ramos D. Fibroblast growth factor 21 and its novel association with oxidative stress. Redox Biology. 2017 Apr 1;11:335-41.

153. Chou WC, Rampanelli E, Li X, Ting JP. Impact of intracellular innate immune receptors on immunometabolism. Cell Mol Immunol. 2022 Mar;19(3):337-351.

154. Kiikka, D. The Basic of Energy production: Aerobic Respiration. The Sports EDU. 2021 Mar 3.

155. Porges SW. Polyvagal Theory: A Science of Safety. Frontiers in Integrative Neuroscience. 2022;16.

156. Spangler JB, Moraga I, Mendoza JL, Garcia KC. Insights into cytokine–receptor interactions from cytokine engineering. Annual Review of Immunology. 2015 Mar 21;33:139.

157. Master RK, Aron LY, Woolf S. CHANGES IN LIFE EXPECTANCY BETWEEN 2019 AND 2021: UNITED STATES AND 19 PEER COUNTRIES. medRxiv. 2022 Jan 1.

158. Black RJ, Friedman RM. Cytokines and oncogene activity. Cancer Surveys. 1989 Jan 1;8(4):725-39.

159. Furman D, Campisi J, Verdin E, Carrera-Bastos P, Targ S, Franceschi C, Ferrucci L, Gilroy DW, Fasano A, Miller GW, Miller AH. Chronic inflammation in the etiology of disease across the life span. Nature Medicine. 2019 Dec;25(12):1822-32.

160. Farooq RK, Asghar K, Kanwal S, Zulqernain A. Role of inflammatory cytokines in depression: Focus on interleukin-1b. Biomedical Reports. 2017 Jan 1;6(1):15-20.

161. Hoge EA, Brandstetter K, Moshier S, Pollack MH, Wong KK, Simon NM. Broad spectrum of cytokine abnormalities in panic disorder and posttraumatic stress disorder. Depression and Anxiety. 2009 May;26(5):447-55.

162. Monji A, Kato T, Kanba S. Cytokines and schizophrenia: Microglia hypothesis of schizophrenia. Psychiatry and Clinical Neurosciences. 2009 Jun;63(3):257-65.

163. Cole SW. Social regulation of human gene expression: mechanisms and implications for public health. American Journal of Public Health. 2013 Oct;103(S1):S84-92.

164. Slavich GM, Way BM, Eisenberger NI, Taylor SE. Neural sensitivity to social rejection is associated with inflammatory responses to social stress. Proceedings of the National Academy of Sciences. 2010 Aug 17;107(33):14817-22.

165. Abboud FM, Harwani SC, Chapleau MW. Autonomic neural regulation of the immune system: implications for hypertension and cardiovascular disease. Hypertension. 2012 Apr;59(4):755-62.

166. Porges SW. The polyvagal theory: Neurophysiological foundations of emotions, attachment, communication, and self-regulation (Norton Series on Interpersonal Neurobiology). WW Norton & Company; 2011 Apr 25.

167. Toufexis D, Rivarola MA, Lara H, Viau V. Stress and the reproductive axis. Journal of Neuroendocrinology. 2014 Sep;26(9):573-86.

168. Bhatia V, Tandon RK. Stress and the gastrointestinal tract. Journal of Gastroenterology and Hepatology. 2005 Mar;20(3):332-9.

169. Morita, W et al Cytokines in tendon disease Bone Joint Res. 2017 Dec; 6(12): 656– 664.

170. Zhao B. TNF and bone remodeling. Current Osteoporosis Reports. 2017 Jun;15(3):126-34.

171. Molnar V, Matišić V, Kodvanj I, Bjelica R, Jeleč Ž, Hudetz D, Rod E, Čukelj F, Vrdoljak T, Vidović D, Staresinić M. Cytokines and chemokines involved in osteoarthritis pathogenesis. International Journal of Molecular sciences. 2021 Aug 26;22(17):9208.

172. De Giorgio R, Barbara G, Furness JB, Tonini M. Novel therapeutic targets for enteric nervous system disorders. Trends in Pharmacological Sciences. 2007 Sep 1;28(9):473-81.

173. Gustafson C. James Gordon, MD: The Potential of Mind-Body Self Care to Free the World From the Effects of Trauma. Integrative Medicine: A Clinician's Journal. 2016 Apr;15(2):54.

174. Dantzer R. Neuroimmune interactions: from the brain to the immune system and vice versa. Physiological Reviews. 2018 Jan 1;98(1):477-504.

175. Esquivel-Rendón E, Vargas-Mireles J, Cuevas-Olguín R, Miranda-Morales M, Acosta-Mares P, García-Oscos F, Pineda JC, Salgado H, Rose-John S, Atzori M. Interleukin 6 dependent synaptic plasticity in a social defeat-susceptible prefrontal cortex circuit. Neuroscience. 2019 Aug 21;414:280-96.

176. Yin W, Gallagher NR, Sawicki CM, McKim DB, Godbout JP, Sheridan JF. Repeated social defeat in female mice induces anxiety-like behavior associated with enhanced myelopoiesis and increased monocyte accumulation in the brain. Brain, Behavior, and Immunity. 2019 May 1;78:131-42.

177. Halaris A. Inflammation and depression but where does the inflammation come from? Current Opinion in Psychiatry. 2019 Sep 1;32(5):422-8.

178. Pandey GN, Rizavi HS, Ren X, Fareed J, Hoppensteadt DA, Roberts RC, Conley RR, Dwivedi Y. Proinflammatory cytokines in the prefrontal cortex of teenage suicide victims. Journal of Psychiatric Research. 2012 Jan 1;46(1):57-63.

179. Mamdani F, Weber MD, Bunney B, Burke K, Cartagena P, Walsh D, Lee FS, Barchas J, Schatzberg AF, Myers RM, Watson SJ. Identification of potential blood biomarkers associated with suicide in major depressive disorder. Translational Psychiatry. 2022 Apr 14;12(1):1-0.

180. Cortese S, Angriman M, Comencini E, Vincenzi B, Maffeis C. Association between inflammatory cytokines and ADHD symptoms in children and adolescents with obesity: A pilot study. Psychiatry Research. 2019 Aug 1;278:7-11.

181. Miller AH, Haroon E, Raison CL, Felger JC. Cytokine targets in the brain: impact on neurotransmitters and neurocircuits. Depression and Anxiety. 2013 Apr;30(4):297-306.

182. Miklowitz DJ, Portnoff LC, Armstrong CC, Keenan-Miller D, Breen EC, Muscatell KA, Eisenberger NI, Irwin MR. Inflammatory cytokines and nuclear factor-kappa B activation in adolescents with bipolar and major depressive disorders. Psychiatry Research. 2016 Jul 30;241:315-22.

183. Wang, Z and Young, R PTSD, a Disorder with an Immunological Component Front Immunol. 2016; 7: 219.

184. Kelmendi B, Adams TG, Yarnell S, Southwick S, Abdallah CG, Krystal JH. PTSD: from neurobiology to pharmacological treatments. European Journal of Psychotraumatology. 2016 Dec 1;7(1):31858.

185. Ashwood P, Krakowiak P, Hertz-Picciotto I, Hansen R, Pessah I, Van de Water J. Elevated plasma cytokines in autism spectrum disorders provide evidence of immune dysfunction and are associated with impaired behavioral outcome. Brain, Behavior, and Immunity. 2011 Jan 1;25(1):40-5.

186. da Silva Araújo T, Chaves Filho AJ, Monte AS, de Góis Queiroz AI, Cordeiro RC, Machado MD, de Freitas Lima R, de Lucena DF, Maes M, Macêdo D. Reversal of schizophrenia-like symptoms and immune alterations in mice by immunomodulatory drugs. Journal of Psychiatric Research. 2017 Jan 1;84:49-58.

187. King S, Holleran L, Mothersill D, Patlola S, Rokita K, McManus R, Kenyon M, McDonald C, Hallahan B, Corvin A, Morris D. Early life Adversity, functional connectivity and cognitive performance in Schizophrenia: The mediating role of IL-6. Brain, Behavior, and Immunity. 2021 Nov 1;98:388-96.

188. Pandey GN, Rizavi HS, Zhang H, Ren X. Abnormal gene and protein expression of inflammatory cytokines in the postmortem brain of schizophrenia patients. Schizophrenia Research. 2018 Feb 1;192:247-54.

189. Tian YE, Di Biase MA, Mosley PE, Lupton MK, Xia Y, Fripp J, Breakspear M, Cropley V, Zalesky A. Evaluation of Brain-Body Health in Individuals With Common Neuropsychiatric Disorders. JAMA Psychiatry. 2023 Apr 26.

190. Hulbert JC, Henson RN, Anderson MC. Inducing amnesia through systemic suppression. Nature Communications. 2016 Mar 15;7(1):1-9.

191. Zuccoli GS, Saia-Cereda VM, Nascimento JM, Martins-de-Souza D. The energy metabolism dysfunction in psychiatric disorders postmortem brains: focus on proteomic evidence. Frontiers in neuroscience. 2017 Sep 7;11:493.

192. Tran, T. Chronic Feelings of Emptiness Increase Emotional Awareness Problems and Depression and Anxiety Levels in Adult Psychiatric Inpatients. Anxiety and Depression Association of America (ADAA) Conference 2022: Abstract 78. Presented 2022, April 14.

193. Lane, R. D., Subic-Wrana, C., Greenberg, L., & Yovel, I. The role of enhanced emotional awareness in promoting change across psychotherapy modalities. The role of enhanced emotional awareness in promoting change across psychotherapy modalities. Journal of Psychotherapy Integration, 2022, 32(2), 131–150.

194. Bachner-Melman R, Watermann Y, Lev-Ari L, Zohar AH. Associations of self-repression with disordered eating and symptoms of other psychopathologies for men and women. Journal of Eating Disorders. 2022 Dec;10(1):1-2.

195. Stibel JM. Climate Change Influences Brain Size in Humans. Brain Behavior and Evolution. 2023 Apr 3;98(2):93-106.

196. Kieling C, Buchweitz C, Caye A, Silvani J, Ameis SH, Brunoni AR, Cost KT, Courtney DB, Georgiades K, Merikangas KR, Henderson JL. Worldwide Prevalence and Disability From Mental Disorders Across Childhood and Adolescence: Evidence From the Global Burden of Disease Study. JAMA psychiatry. 2024 Jan 31.

197. Colagrossi, M. Ten reasons Why the Finnish School System is the Best in The World, World economic Summit, 2018, Sep 10.

198. Herrmann-Lingen C. Mental disorders and cardiovascular disease: not just an issue of older age. European Journal of Preventive Cardiology. 2023 May 8:zwad118.

199. Zeighami Y, Bakken TE, Nickl-Jockschat T, Peterson Z, Jegga AG, Miller JA, Schulkin J, Evans AC, Lein ES, Hawrylycz M. A comparison of anatomic and cellular transcriptome structures across 40 human brain diseases. Plos Biology. 2023 Apr 20;21(4):e3002058.

200. Fragkaki I, Verhagen M, van Herwaarden AE, Cima M. Daily oxytocin patterns in relation to psychopathy and childhood trauma in residential youth. Psychoneuroendocrinology. 2019 Apr 1;102:105-13.

201. Grafman J. Anger is a wind that blows out the light of the mind (old proverb). Molecular Psychiatry. 2003 Feb 1;8(2):131-.

202. Schiltz K, Witzel JG, Bausch-Hölterhoff J, Bogerts B. High prevalence of brain pathology in violent prisoners: a qualitative CT and MRI scan study. European Archives of Psychiatry and Clinical Neuroscience. 2013 Oct;263(7):607-16.

203. Ly M, Motzkin JC, Philippi CL, Kirk GR, Newman JP, Kiehl KA, Koenigs M. Cortical thinning in psychopathy. American Journal of Psychiatry. 2012 Jul;169(7):743-9.

204. Lawrence WR, Freedman ND, McGee-Avila JK, et al. Trends in Mortality From Poisonings, Firearms, and All Other Injuries by Intent in the U.S., 1999-2020. *JAMA Intern Med.* Published online July 03, 2023.

205. Barron-Lopez, L, Lane, S. Most Young Americans Feel Unsafe and Support Stricter Gun Laws, New Survey Shows. PBS News Hour. Jul 24 2023.

206. Loeber R, Pardini D, Homish DL, Wei EH, Crawford AM, Farrington DP, Stouthamer-Loeber M, Creemers J, Koehler SA, Rosenfeld R. The prediction of violence and homicide in young men. Journal of Consulting and Clinical Psychology. 2005 Dec;73(6):1074.

207. Sterbenz C. Why Norway's prison system is so successful. Business Insider. 2014 Dec 11;11.

208. Achur, R Circulating Cytokines as Biomarkers of Alcohol Abuse and Alcoholism J Neuroimmune Pharmacol. 2010 Mar; 5(1): 83–91.

209. Joutsa J, Moussawi K, Siddiqi SH, Abdolahi A, Drew W, Cohen AL, Ross TJ, Deshpande HU, Wang HZ, Bruss J, Stein EA. Brain lesions disrupting addiction map to a common human brain circuit. Nature Medicine. 2022 Jun 13:1-7.

210. Hofford RS, Russo SJ, Kiraly DD. Neuroimmune mechanisms of psychostimulant and opioid use disorders. European Journal of Neuroscience. 2019 Aug;50(3):2562-73.

211. Lin AL, Nah G, Tang JJ, Vittinghoff E, Dewland TA, Marcus GM. Cannabis, cocaine, methamphetamine, and opiates increase the risk of incident atrial fibrillation. European Heart Journal. 2022 Dec 14;43(47):4933-42.

212. Jiménez-González A, Gómez-Acevedo C, Ochoa-Aguilar A, Chavarría A. The Role of Glia in Addiction: Dopamine as a Modulator of Glial Responses in Addiction. Cellular and Molecular Neurobiology. 2021 May 31:1-2.

213. DRUG DECRIMINALISATION IN PORTUGAL: SETTING THE RECORD STRAIGHT, Transform Drug Policy Foundation, 2021 May 13.

214. Duca LM, Helmick CG, Barbour KE, Nahin RL, Von Korff M, Murphy LB, Theis K, Guglielmo D, Dahlhamer J, Porter L, Falasinnu T. A review of potential national chronic pain surveillance systems in the United States. The Journal of Pain. 2022 Sep 1;23(9):1492-509.

215. Zhang K, Sun J, Zhang Q, Zhang J, He L, Wang Z, Hu L. The association between childhood trauma and pain symptoms in depressed adults: the moderating role of anxious attachment. Clinical Psychology & Psychotherapy. 2023 Jan 11.

216. Yong RJ, Mullins PM, Bhattacharyya N. Prevalence of chronic pain among adults in the United States. Pain. 2022 Feb 1;163(2):e328-32.

217. Rogoz K, Andersen HH, Stjarne L, Kullander K, Lagerstrom MC. Glutamate, Substance P and CGRP cooperate in inflammation-induced heat hyperalgesia. Molecular Pharmacology. 2013 Nov 25.

218. Apkarian AV, Baliki MN, Geha PY. Towards a theory of chronic pain. Progress in Neurobiology. 2009 Feb 1;87(2):81-97.

219. Ong WY, Stohler CS, Herr DR. Role of the prefrontal cortex in pain processing. Molecular Neurobiology. 2019 Feb;56(2):1137-66.

220. Yang S, Chang MC. Chronic pain: structural and functional changes in brain structures and associated negative affective states. International Journal of Molecular Sciences. 2019 Jun 26;20(13):3130.

221. Stahl SM. Fibromyalgia—pathways and neurotransmitters. Human Psychopharmacology: Clinical and Experimental. 2009 Jun;24(S1):S11-7.

222. Jin K, Lu J, Yu Z, Shen Z, Li H, Mou T, Xu Y, Huang M. Linking peripheral IL-6, IL-1b and hypocretin-1 with cognitive impairment from major depression. Journal of Affective Disorders. 2020 Dec 1;277:204-11.

223. Pohóczky K, Kun J, Szentes N, Aczél T, Urbán P, Gyenesei A, Szőke É, Sensi S, Dénes Á, Goebel A, Tékus V. Discovery of novel targets in a complex regional pain syndrome mouse model by transcriptomics: TNF and JAK-STAT pathways. Pharmacological Research. 2022 Aug 1;182:106347.

224. Harden RN, McCabe CS, Goebel A, Massey M, Suvar T, Grieve S, Bruehl S. Complex Regional Pain Syndrome: Practical Diagnostic and Treatment Guidelines. Pain Medicine. 2022 May;23(Supplement_1):S1-53.

225. Stanton-Hicks M. Complex regional pain syndrome. Clinical Pain Management: A Practical Guide. 2022 Mar 9:381-95.

226. Ashar YK, Gordon A, Schubiner H, Uipi C, Knight K, Anderson Z, Carlisle J, Polisky L, Geuter S, Flood TF, Kragel PA. Effect of pain reprocessing therapy vs placebo and usual care for patients with chronic back pain: a randomized clinical trial. JAMA Psychiatry. 2022 Jan 1;79(1):13-23.

227. Abbass A, Town JM, Driessen E. Intensive short-term dynamic psychotherapy: A treatment overview and empirical basis. Research in Psychotherapy: Psychopathology, Process and Outcome. 2013 Aug 17;16(1):6-15.

228. Cooper A, Abbass A, Zed J, Bedford L, Sampalli T, Town J. Implementing a psychotherapy service for medically unexplained symptoms in a primary care setting. Journal of Clinical Medicine. 2017 Nov 29;6(12):109.

229. Sullivan MD, Ballantyne JC. Randomised trial reveals opioids relieve acute back pain no better than placebo. The Lancet. 2023 Jun 27.

230. Owolabi MO, Leonardi M, Bassetti C, Jaarsma J, Hawrot T, Makanjuola AI, Dhamija RK, Feng W, Straub V, Camaradou J, Dodick DW. Global synergistic actions to improve brain health for human development. Nature Reviews Neurology. 2023 May 19:1-3.

231. Congress of the European Academy of Neurology (EAN) 2023: Abstract EPO-236. Presented July 2, 2023.

232. Elser H, Horváth-Puhó E, Gradus JL, Smith ML, Lash TL, Glymour MM, Sørensen HT, Henderson VW. Association of Early-, Middle-, and Late-Life Depression With Incident Dementia in a Danish Cohort. JAMA neurology. 2023 Jul 24.

233. Halahakoon DC, Kieslich K, O'Driscoll C, Nair A, Lewis G, Roiser JP. Reward-processing behavior in depressed participants relative to healthy volunteers: A systematic review and meta-analysis. JAMA Psychiatry. 2020 Dec 1;77(12):1286-95.

234. Shen C, Rolls E, Cheng W, Kang J, Dong G, Xie C, Zhao XM, Sahakian B, Feng J. Associations of Social Isolation and Loneliness With Later Dementia. Neurology. 2022 Jun 8.

235. Novellino F, Saccà V, Donato A, Zaffino P, Spadea MF, Vismara M, Arcidiacono B, Malara N, Presta I, Donato G. Innate immunity: a common denominator between neurodegenerative and neuropsychiatric diseases. International Journal of Molecular sciences. 2020 Feb 7;21(3):1115.

236. Tani M, Akashi N, Hori K, Konishi K, Kitajima Y, Tomioka H, Inamoto A, Hirata A, Tomita A, Koganemaru T, Takahashi A. Anticholinergic activity and schizophrenia. Neurodegenerative Diseases. 2015;15(3):168-74.

237. Milligan Armstrong A, Porter T, Quek H, White A, Haynes J, Jackaman C, Villemagne V, Munyard K, Laws SM, Verdile G, Groth D. Chronic stress and Alzheimer's disease: the interplay between the hypothalamic–pituitary–adrenal axis, genetics and microglia. Biological Reviews. 2021 Oct;96(5):2209-28.

238. Stamouli EC, Politis AM. Pro-inflammatory cytokines in Alzheimer's disease. Psychiatrike=Psychiatriki. 2016 Oct 1;27(4):264-75.

239. Tzeng TC, Golenbock D. NLRP3 inflammasome activation in Alzheimer's disease (INC9P. 446). The Journal of Immunology. 2014 May 1;192(1_Supplement):188-5.

240. Willette AA, Bendlin BB, McLaren DG, Canu E, Kastman EK, Kosmatka KJ, Xu G, Field AS, Alexander AL, Colman RJ, Weindruch RH. Age-related changes in neural volume and microstructure associated with interleukin-6 are ameliorated by a calorie-restricted diet in old rhesus monkeys. Neuroimage. 2010 Jul 1;51(3):987-94.

241. Lee JH, Yang DS, Goulbourne CN, Im E, Stavrides P, Pensalfini A, Chan H, Bouchet-Marquis C, Bleiwas C, Berg MJ, Huo C. Faulty autolysosome acidification in Alzheimer's disease mouse models induces autophagic build-up of Ab in neurons, yielding senile plaques. Nature Neuroscience. 2022 Jun 2:1-4.

242. Sachin P, Gadani S, Cronk J, Norris G, Kipnis J. Interleukin-4: a cytokine to remember. J Immunol. 2012;189:4213-421.

243. Reale M, Iarlori C, Thomas A, Gambi D, Perfetti B, Di Nicola M, Onofrj M. Peripheral cytokines profile in Parkinson's disease. Brain, Behavior, and Immunity. 2009 Jan 1;23(1):55-63.

244. Porro C, Cianciulli A, Panaro MA. The regulatory role of IL-10 in neurodegenerative diseases. Biomolecules. 2020 Jul 9;10(7):1017.

245. Otaiku AI. Distressing dreams and risk of Parkinson's disease: a population-based cohort study. EClinicalMedicine. 2022 Jun 1;48.

246. Jones MB, Gates R, Gibson L, Broadway D, Bhatti G, Tea J, Guerra A, Li R, Varman B, Elammari M, Jorge RE. Post-traumatic stress disorder and risk of degenerative synucleinopathies: systematic review and meta-analysis. The American Journal of Geriatric Psychiatry. 2023 May 3.

247. Walker KA, Chen J, Shi L, Yang Y, Fornage M, Zhou L, Schlosser P, Surapaneni A, Grams ME, Duggan MR, Peng Z, Gomez GT, Tin A, Hoogeveen RC, Sullivan KJ, Ganz P, Lindbohm JV, Kivimaki M, Nevado-Holgado AJ, Buckley N, Gottesman RF, Mosley TH, Boerwinkle E, Ballantyne CM, Coresh J. Proteomics analysis of plasma from middle-aged adults identifies protein markers of dementia risk in later life. Sci Transl Med. 2023 Jul 19;15(705):eadf5681.

248. Kang J, Eun Y, Jang W, Cho MH, Han K, Jung J, Kim Y, Kim GT, Shin DW, Kim H. Rheumatoid Arthritis and Risk of Parkinson Disease in Korea. JAMA Neurology. 2023 May 1.

249. Livingston G, Huntley J, Sommerlad A, Ames D, Ballard C, Banerjee S, Brayne C, Burns A, Cohen-Mansfield J, Cooper C, Costafreda SG. Dementia prevention, intervention, and care: 2020 report of the Lancet Commission. The Lancet. 2020 Aug 8;396(10248):413-46.

250. Ihara K, Oguro A, Imaishi H. Diagnosis of Parkinson's disease by investigating the inhibitory effect of serum components on P450 inhibition assay. Scientific Reports. 2022 Apr 22;12(1):1.

251. Jones SV, Kounatidis I. Nuclear factor-kappa B and Alzheimer disease, unifying genetic and environmental risk factors from cell to humans. Frontiers in Immunology. 2017 Dec 11;8:1805.

252. Zhou RZ, Vetrano DL, Grande G, Duell F, Jönsson L, Laukka EJ, Fredolini C, Winblad B, Tjernberg L, Schedin-Weiss S. A glycan epitope correlates with tau in serum and predicts progression to Alzheimer's disease in combination with APOE4 allele status. Alzheimer's & Dementia. 2023 Apr 12.

253. Festa BP, Siddiqi FH, Jimenez-Sanchez M, Won H, Rob M, Djajadikerta A, Stamatakou E, Rubinsztein DC. Microglial-to-neuronal CCR5 signaling regulates autophagy in neurodegeneration. Neuron. 2023 Apr 18.

254. Mirhafez SR, Pasdar A, Avan A, Esmaily H, Moezzi A, Mohebati M, Meshkat Z, Mehrad-Majd H, Eslami S, Rahimi HR, Ghazavi H. Cytokine and growth factor profiling in patients with the metabolic syndrome. British Journal of Nutrition. 2015 Jun;113(12):1911-9.

255. Kim TN, Choi KM. The implications of sarcopenia and sarcopenic obesity on cardiometabolic disease. Journal of Cellular Biochemistry. 2015 Jul;116(7):1171-8.

256. Ong KL, Stafford LK, McLaughlin SA, Boyko EJ, Vollset SE, Smith AE, Dalton BE, Duprey J, Cruz JA, Hagins H, Lindstedt PA. Global, regional, and national burden of diabetes from 1990 to 2021, with projections of prevalence to 2050: a systematic analysis for the Global Burden of Disease Study 2021. The Lancet. 2023 Jun 22.

257. Misiak B, Stańczykiewicz B, Pawlak A, Szewczuk-Bogusławska M, Samochowiec J, Samochowiec A, Tyburski E, Juster RP. Adverse childhood experiences and low socioeconomic status with respect to allostatic load in adulthood: a systematic review. Psychoneuroendocrinology. 2022 Feb 1;136:105602.

258. Xia W, Veeragandham P, Cao Y, Xu Y, Rhyne TE, Qian J, Hung CW, Zhao P, Jones Y, Gao H, Liddle C. Obesity causes mitochondrial fragmentation and dysfunction in white adipocytes due to RalA activation. Nature Metabolism. 2024 Jan 29:1-7.

259. Aminian A, Wilson R, Al-Kurd A, Tu C, Milinovich A, Kroh M, Rosenthal RJ, Brethauer SA, Schauer PR, Kattan MW, Brown JC. Association of Bariatric Surgery With Cancer Risk and Mortality in Adults With Obesity. JAMA. 2022.

260. Sylow L, Grand MK, von Heymann A, Persson F, Siersma V, Kriegbaum M, Lykkegaard Andersen C, Johansen C. Incidence of New-Onset Type 2 Diabetes After Cancer: A Danish Cohort Study. Diabetes Care. 2022 Jun 2;45(6):e105-6.

261. Johnson RJ, Tolan DR, Bredesen D, Nagel M, Sánchez-Lozada LG, Fini M, Burtis S, Lanaspa MA, Perlmutter D. Could Alzheimer's Disease Be a Maladaptation of an Evolutionary Survival Pathway Mediated by Intracerebral Fructose and Uric acid Metabolism?. The American Journal of Clinical Nutrition. 2023 Jan 11.

262. Neth BJ, Craft S. Insulin resistance and Alzheimer's disease: bioenergetic linkages. Frontiers in Aging Neuroscience. 2017 Oct 31;9:345.

263. Wing RR, Phelan S. Long-term weight loss maintenance–. The American Journal of Clinical Nutrition. 2005 Jul 1;82(1):222S-5S

264. Armstrong A, Jungbluth Rodriguez K, Sabag A, Mavros Y, Parker HM, Keating SE, Johnson NA. Effect of aerobic exercise on waist circumference in adults with overweight or obesity: A systematic review and meta-analysis. Obesity Reviews. 2022 Aug;23(8):e13446

265. Savchenko LG, Digtiar NI, Selikhova LG, Kaidasheva EI, Shlykova OA, Vesnina LE, Kaidashev IP. Liraglutide exerts an anti-inflammatory action in obese patients with type 2 diabetes. Romanian Journal of Internal Medicine. 2019 Sep 1;57(3):233-40.

266. Rizzo M, Nauck MA, Mantzoros CS. Incretin-based therapies in 2021–Current status and perspectives for the future. Metabolism-Clinical and Experimental. 2021 Sep 1;122.

267. Sanchez-Muñoz F, Dominguez-Lopez A, Yamamoto-Furusho JK. Role of cytokines in inflammatory bowel disease. World Journal of Gastroenterology: WJG. 2008 Jul 7;14(27):4280.

268. Bonaz B, Bazin T, Pellissier S. The vagus nerve at the interface of the microbiota-Gut-Brain Axis. Front. Neurosci. Frontiers Media SA. 2018;12.

269. O'Reardon JP, Cristancho P, Peshek AD. Vagus nerve stimulation (VNS) and treatment of depression: to the brainstem and beyond. Psychiatry (Edgmont). 2006 May 1;3(5):54.

270. Ben-Menachem E. Vagus-nerve stimulation for the treatment of epilepsy. The Lancet Neurology. 2002 Dec 1;1(8):477-82.

271. Ma, C Wang,D Snyder,H and Griffin, P Alzheimer's Association International Conference (AAIC) 2023: Abstract 73719. Presented July 19, 2023.

272. Kelly MJ, Breathnach C, Tracey KJ, Donnelly SC. Manipulation of the inflammatory reflex as a therapeutic strategy. Cell Reports Medicine. 2022 Jul 19;3(7):100696.

273. Tracey KJ. The inflammatory reflex. Nature. 2002 Dec;420(6917):853-9.

274. Schneider KM, Blank N, Alvarez Y, Thum K, Lundgren P, Litichevskiy L, Sleeman M, Bahnsen K, Kim J, Kardo S, Patel S. The enteric nervous system relays psychological stress to intestinal inflammation. Cell. 2023 May 25.

275. López-Pulido EI, Medrano-González ID, Becerra-Ruiz JS, Alonso-Sanchez CC, Vázquez-Jiménez SI, Guerrero-Velázquez C, Guzmán-Flores JM. Alteration of cytokines in saliva of children with caries and obesity. Odontology. 2021 Jan;109(1):11-7.

276. Tawfig N. Proinflammatory cytokines and periodontal disease. Journal of Dental Problems and Solutions. 2016 Apr 5;3(1):012-7.

277. Bacali C, Vulturar R, Buduru S, Cozma A, Fodor A, Chiș A, Lucaciu O, Damian L, Moldovan ML. Oral Microbiome: Getting to Know and Befriend Neighbors, a Biological Approach. Biomedicines. 2022 Mar 14;10(3):671.

278. Ball J, Darby I. Mental health and periodontal and peri-implant diseases. Periodontology 2000. 2022 Aug 1.

279. Lamont RJ, Jenkinson HF. Oral microbiology at a glance. John Wiley & Sons; 2010 Feb 22.

280. Williams JW, Huang LH, Randolph GJ. Cytokine circuits in cardiovascular disease. Immunity. 2019 Apr 16;50(4):941-54.

281. Kennedy KG, Islam AH, Karthikeyan S, Metcalfe AW, McCrindle BW, MacIntosh BJ, Black S, Goldstein BI. Differential association of endothelial function with brain structure in youth with versus without bipolar disorder. Journal of Psychosomatic Research. 2023 Apr 1;167:111180.

282. Tian Y, Ullah H, Gu J, Li K. Immune-metabolic mechanisms of post-traumatic stress disorder and atherosclerosis. Frontiers in Physiology. 2023 Feb 8;14:180.

283. Agbaje AO. Arterial stiffness preceding metabolic syndrome in 3,862 adolescents: a mediation and temporal causal longitudinal birth cohort study. American Journal of Physiology-Heart and Circulatory Physiology. 2023 Jun 1;324(6):H905-11.

284. Rhee TM, Choi J, Choi EK, Lee KY, Ahn HJ, Kwon S, Lee SR, Oh S, Lip G. CE-452780-4 NEUROTICISM AND THE OCCURRENCE OF ATRIAL FIBRILLATION: EVIDENCE FROM OBSERVATIONAL EPIDEMIOLOGIC AND MENDELIAN RANDOMIZATION. Heart Rhythm. 2023 May 1;20(5):S56.

285. Gulhar R, Ashraf MA, Jialal I. Physiology, acute phase reactants. 2018.

286. Duewell P, Kono H, Rayner KJ, Sirois CM, Vladimer G, Bauernfeind FG, Abela GS, Franchi L, Nuñez G, Schnurr M, Espevik T. NLRP3 inflammasomes are required for atherogenesis and activated by cholesterol crystals. Nature. 2010 Apr;464(7293):1357-61.

287. Tkachev A, Stekolshchikova E, Vanyushkina A, Zhang H, Morozova A, Zozulya S, Kurochkin I, Anikanov N, Egorova A, Yushina E, Vogl T. Lipid Alteration Signature in the Blood Plasma of Individuals With Schizophrenia, Depression, and Bipolar Disorder. JAMA Psychiatry. 2023 Jan 25.

288. Moser ED, Manemann SM, Larson NB, Sauver JL, Takahashi PY, Mielke MM, Rocca WA, Olson JE, Roger VL, Remaley AT, Decker PA. Association Between Fluctuations in Blood Lipid Levels Over Time With Incident Alzheimer Disease and Alzheimer Disease Related Dementias. Neurology. 2023 Jul 5.

289. Berliner JA, Watson AD. A role for oxidized phospholipids in atherosclerosis. New England Journal of Medicine. 2005 Jul 7;353(1):9-11.

290. Dou H, Kotini A, Liu W, Fidler T, Endo-Umeda K, Sun X, Olszewska M, Xiao T, Abramowicz S, Yalcinkaya M, Hardaway B. Oxidized phospholipids promote NETosis and arterial thrombosis. Circulation. 2021 Dec 14;144(24):1940-54.

291. Lotfollahi Z, Dawson J, Fitridge R, Bursill C. The anti-inflammatory and proangiogenic properties of high-density lipoproteins: an emerging role in diabetic wound healing. Advances in Wound Care. 2021 Jul 1;10(7):370-80.

292. West AP, Shadel GS. Mitochondrial DNA in innate immune responses and inflammatory pathology. Nature Reviews Immunology. 2017 Jun;17(6):363-75.

293. S, Tran VH. Postural Orthostatic Tachycardia Syndrome.[Updated 2021 Aug 11]. In: StatPearls[Internet]. Treasure Island (FL): StatPearls Publishing; 2022 Jan-.

294. Dahan, S., Tomljenovic, L., & Shoenfeld, Y. (2016). Postural Orthostatic Tachycardia Syndrome (POTS)--A novel member of the autoimmune family. *Lupus*, 25(4), 339–342.

295. Gunning III WT, Stepkowski SM, Kramer PM, Karabin BL, Grubb BP. Inflammatory biomarkers in postural orthostatic tachycardia syndrome

with elevated G-protein-coupled receptor autoantibodies. Journal of Clinical Medicine. 2021 Feb 6;10(4):623.

296. Amorim MR, de Deus JL, Cazuza RA, Mota C, da Silva LE, Borges GS, Batalhão ME, Cárnio EC, Branco LG. Neuroinflammation in the NTS is associated with changes in cardiovascular reflexes during systemic inflammation. Journal of Neuroinflammation. 2019 Dec;16(1):1-5.

297. Singh S, Desai R, Gandhi Z, Fong HK, Doreswamy S, Desai V, Chockalingam A, Mehta PK, Sachdeva R, Kumar G. Takotsubo syndrome in patients with COVID-19: a systematic review of published cases. SN Comprehensive Clinical Medicine. 2020 Nov;2:2102-8.

298. Ye M, Martinez A, Hessler D, De La Rosa R, Long D, Thakur N. Associations Between Food Insecurity and Assistance With Asthma and Symptomology Burden Among Children in a Safety-net Practice. InA22. UNDERSTANDING THE IMPACT OF HEALTH DISPARITIES ON LUNG HEALTH 2023 May (pp. A1097-A1097). American Thoracic Society.

299. Sharma R, Humphrey JL, Frueh L, Kinnee EJ, Sheffield PE, Clougherty JE. Neighborhood violence and socioeconomic deprivation influence associations between acute air pollution and temperature on childhood asthma in New York City. Environmental Research. 2023 May 25:116235.

300. Zhang, P., Carleton, T., Lin, L. *et al.* Estimating the role of air quality improvements in the decline of suicide rates in China. *Nat Sustain* (2024). https://doi.org/10.1038/s41893-024-01281-2

301. Skeen E, Moore CM, Liu AH, Seibold MA, Hamlington KL. Neighborhood-level Child Opportunity Predicts Exacerbation-prone Status in a Cohort of Urban Children With Asthma. InB96. BEST IN PEDIATRICS 2023 May (pp. A4227-A4227). American Thoracic Society.

302. Cardet JC, Bulkhi AA, Lockey RF. Nonrespiratory comorbidities in asthma. The Journal of Allergy and Clinical Immunology: In Practice. 2021 Nov 1;9(11):3887-97.

303. Guo Y, Bian J, Chen Z, Fishe JN, Zhang D, Braithwaite D, George TJ, Shenkman EA, Licht JD. Cancer incidence after asthma diagnosis: Evidence from a large clinical research network in the United States. Cancer Medicine. 2023 Mar 31.

304. Lambrecht BN, Hammad H, Fahy JV. The cytokines of asthma. Immunity. 2019 Apr 16;50(4):975-91.

305. Margolis RH, Patel SJ, Krueger J, Brewer T, Williams A, Stringfield S, Teach SJ, Parikh K. Association between social needs and asthma control among children evaluated at a single-center high-risk asthma clinic. The Journal of Allergy and Clinical Immunology: In Practice. 2023 Mar 13.

306. Schwartz U, Llamazares Prada M, Pohl ST, Richter M, Tamas R, Schuler M, Keller C, Mijosek V, Muley T, Schneider MA, Quast K. High-resolution transcriptomic and epigenetic profiling identifies novel regulators of COPD. The EMBO Journal. 2023 Mar 30:e111272.

307. Borthwick LA, Wynn TA, Fisher AJ. Cytokine mediated tissue fibrosis. Biochimica et Biophysica Acta (BBA 2013 Jul 1;1832(7):1049-60.

308. Yadav P, Sharma A, Singh L, Gupta R. Management of fibrocystic breast disease: A comprehensive review. Journal of Advanced Scientific Research. 2020 Nov 10;11(04):30-7.

309. Karande S, Sharma K, Kumar A, Charan S, Patil C, Sharma A. Potential Role of Biopeptides in the Treatment of Idiopathic Pulmonary Fibrosis. Health Sciences Review. 2023 Feb 8:100081.

310. Gewiss C, Augustin M. Recent insights into comorbidities in atopic dermatitis. Expert Review of Clinical Immunology. 2023 Apr 3;19(4):393-404.

311. Guan Q, Gao X, Wang J, Sun Y, Shekhar S. Cytokines in autoimmune disease. Mediators of Inflammation. 2017 Jul 11;2017.

312. Zhang YR, Yang L, Wang HF, Wu BS, Huang SY, Cheng W, Feng JF, Yu JT. Immune-mediated diseases are associated with a higher incidence of dementia: a prospective cohort study of 375,894 individuals. Alzheimers Res Ther. 2022 Sep 13;14(1):130.

313. Dube SR, Fairweather D, Pearson WS, Felitti VJ, Anda RF, Croft JB. Cumulative childhood stress and autoimmune diseases in adults. Psychosomatic Medicine. 2009 Feb;71(2):243.

314. Shaw MT, Pawlak NO, Frontario A, Sherman K, Krupp LB, Charvet LE. Adverse childhood experiences are linked to age of onset and reading recognition in multiple sclerosis. Frontiers in Neurology. 2017 Jun 2;8:242.

315. Polick CS, Ploutz-Snyder R, Braley TJ, Connell CM, Stoddard SA. Associations among stressors across the lifespan, disability, and relapses in adults with multiple sclerosis. Brain and Behavior. 2023 May 21:e3073.

316. van den Bosch AM, Hümmert S, Steyer A, Ruhwedel T, Hamann J, Smolders J, Nave KA, Stadelmann C, Kole MH, Möbius W, Huitinga I. Ultrastructural axon–myelin unit alterations in multiple sclerosis correlate with inflammation. Annals of Neurology. 2023 Apr;93(4):856-70.

317. Paolo Mariotti, Chronic migraine-like headache caused by a demyelinating lesion in the brain stem Pain Med. 2012 Apr;13(4):610-2. doi:10.1111/j.1526-4637.2012.01344.x.

318. Sotzny F, Filgueiras IS, Kedor C, Freitag H, Wittke K, Bauer S, Sepúlveda N, Mathias da Fonseca DL, Baiocchi GC, Marques AH, Kim M. Dysregulated autoantibodies targeting vaso-and immunoregulatory receptors in Post COVID Syndrome correlate with symptom severity. Frontiers in Immunology. 2022:5182.

319. Zia S, Hammond BP, Zirngibl M, Sizov A, Baaklini CS, Panda SP, Ho MF, Lee KV, Mainali A, Burr MK, Williams S. Single-cell microglial transcriptomics during demyelination defines a microglial state required for lytic carcass clearance. Molecular Neurodegeneration. 2022 Dec;17(1):1-24.

320. Kapate N, Dunne M, Kumbhojkar N, Prakash S, Wang LL, Graveline A, Park KS, Chandran Suja V, Goyal J, Clegg JR, Mitragotri S. A backpack-based myeloid cell therapy for multiple sclerosis. Proceedings of the National Academy of Sciences. 2023 Apr 25;120(17):e2221535120.

321. Ghosh T, Almeida R, Zhao C, Mannioui A, Martin E, Fleet A, Chen C, Assinck P, Ellams S, Gonzalez G, Graham S. A retroviral link to vertebrate myelination through retrotransposon RNA-mediated control of myelin gene expression.

322. Shabani Z. Demyelination as a result of an immune response in patients with COVID-19. Acta Neurologica Belgica. 2021 Aug;121(4):859-66.

323. Pisetsky DS. Immune phenotypes in individuals positive for antinuclear antibodies: The impact of race and ethnicity. Journal of Allergy and Clinical Immunology. 2020 Dec 1;146(6):1346-8.

324. Gavrilov YV, Alekseeva TM, Kreis OA, Valko PO, Weber KP, Valko Y. Depression in myasthenia gravis: a heterogeneous and intriguing entity. Journal of Neurology. 2020 Jun;267(6):1802-11.

325. Muri J, Cecchinato V, Cavalli A, Shanbhag AA, Matkovic M, Biggiogero M, Maida PA, Moritz J, Toscano C, Ghovehoud E, Furlan R. Autoantibodies against chemokines post-SARS-CoV-2 infection correlate with disease course. Nature Immunology. 2023 Mar 6:1-8.

326. Kraus A, Buckley KM, Salinas I. Sensing the world and its dangers: An evolutionary perspective in neuroimmunology. Elife. 2021 Apr 26;10:e66706. doi: 10.7554/eLife.66706. PMID: 33900197; PMCID: PMC8075586.

327. Fredrickson BL, Grewen KM, Coffey KA, Algoe SB, Firestine AM, Arevalo JM, Ma J, Cole SW. A functional genomic perspective on human well-being. Proceedings of the National Academy of Sciences. 2013 Aug 13;110(33):13684-9.

328. Balint,E, Feng, E., Giles e.c., et al. Bystander activated CD8+ cells mediate neuropathology during viral infection via antigen-independent cytotoxicity. Nat Commub 15, 896 (2024).

329. Yipp BG, Kubes P. NETosis: how vital is it? Blood, The Journal of the American Society of Hematology. 2013 Oct 17;122(16):2784-94.

330. Moore JX, Andrzejak SE, Bevel MS, Jones SR, Tingen MS. Exploring racial disparities on the association between allostatic load and cancer mortality: A retrospective cohort analysis of NHANES, 1988 through 2019. SSM-Population Health. 2022 Sep 1;19:101185.

331. Donghao Lu Psychiatric disorders, stress increase risk for cervical cancer mortality, Hem-Onc Today August 09, 2019.

332. Lan T, Chen L, Wei X. Inflammatory cytokines in cancer: comprehensive understanding and clinical progress in gene therapy. Cells. 2021 Jan 8;10(1):100.

333. Wolf MM, Rathmell WK, de Cubas AA. Immunogenicity in renal cell carcinoma: shifting focus to alternative sources of tumour-specific antigens. Nature Reviews Nephrology. 2023 Mar 27:1-1.

334. Mosaddeghi P, Farahmandnejad M, Zarshenas MM. The role of transposable elements in aging and cancer. Biogerontology. 2023 Apr 5:1-3.

335. Fardi M, Solali S, Hagh MF. Epigenetic mechanisms as a new approach in cancer treatment: An updated review. Genes & Diseases. 2018 Dec 1;5(4):304-11.

336. Dennis KL, Blatner NR, Gounari F, Khazaie K. Current status of IL-10 and regulatory T-cells in cancer. Curr Opin Oncol. 2013;25:637-45.

337. Liu Z, Liu G, Ha DP, Wang J, Xiong M, Lee AS. ER chaperone GRP78/BiP translocates to the nucleus under stress and acts as a transcriptional regulator. Proceedings of the National Academy of Sciences. 2023 Aug 1;120(31):e2303448120.

338. Hill W, Lim EL, Weeden CE, Lee C, Augustine M, Chen K, Kuan FC, Marongiu F, Evans Jr EJ, Moore DA, Rodrigues FS. Lung adenocarcinoma promotion by air pollutants. Nature. 2023 Apr 6;616(7955):159-67.

339. Sun Y, Guo Y, Shi X, Chen X, Feng W, Wu LL, Zhang J, Yu S, Wang Y, Shi Y. An Overview: The Diversified Role of Mitochondria in Cancer Metabolism. International Journal of Biological Sciences. 2023;19(3):897.

340. Shah, N.M., Jang, H.J., Liang, Y. *et al.* Pan-cancer analysis identifies tumor-specific antigens derived from transposable elements. *Nat Genet* 55, 631–639 (2023).

341. Wang DH. Extrachromosomal DNA appears before cancer forms. Nature. 2023.

342. Cisneros-Villanueva, M., Hidalgo-Pérez, L., Rios-Romero, M. et al. Cell-free DNA analysis in current cancer clinical trials: a review. Br J Cancer 126, 391–400 (2022).

343. Park, WY., Gray, J.M., Holewinski, R.J. *et al.* Apoptosis-induced nuclear expulsion in tumor cells drives S100a4-mediated metastatic outgrowth through the RAGE pathway. *Nat Cancer* 4, 419–435 (2023).

344. Sun Y, Guo Y, Shi X, Chen X, Feng W, Wu LL, Zhang J, Yu S, Wang Y, Shi Y. An Overview: The Diversified Role of Mitochondria in Cancer Metabolism. International Journal of Biological Sciences. 2023;19(3):897.

345. Marín-Hernández A, Gallardo-Pérez JC, Rodríguez-Enríquez S, Encalada R, Moreno-Sánchez R, Saavedra E. Modeling cancer glycolysis. Biochimica et Biophysica Acta (BBA)-Bioenergetics.

346. Kandi V. Bacterial capsule, colony morphology, functions, and its relation to virulence and diagnosis. Annals of Tropical Medicine and Public Health. 2015 Jul 1;8(4):151.

347. Dube DH, Bertozzi CR. Glycans in cancer and inflammation—potential for therapeutics and diagnostics. Nature reviews Drug discovery. 2005 Jun;4(6):477-88.

348. Wood KB, Comba A, Motsch S, Grigera TS, Lowenstein PR. Scale-free correlations and potential criticality in weakly ordered populations of brain cancer cells. Science Advances. 2023 Jun 28;9(26):eadf7170.

349. Cui X, Zhang H, An'na Cao LC, Hu X. Cytokine TNF-a promotes invasion and metastasis of gastric cancer by down-regulating Pentraxin3. Journal of Cancer. 2020;11(7):1800.

350. Pfeffer C. SinghATK. Apoptosis: A Target for Anticancer Therapy. Int J Mol Sci 2018 19.;448.

351. Das D, Karthik N, Taneja R. Crosstalk Between Inflammatory Signaling and Methylation in Cancer. Frontiers in Cell and Developmental Biology. 2021;9.

352. Bhoi A, Yadu B, Chandra J, Keshavkant S. Mutagenesis: A coherent technique to develop biotic stress resistant plants. Plant Stress. 2022 Jan 1;3:100053.

353. Pomella S, Danielli SG, Alaggio R, Breunis WB, Hamed E, Selfe J, Wachtel M, Walters ZS, Schäfer BW, Rota R, Shipley JM. Genomic and Epigenetic Changes Drive Aberrant Skeletal Muscle Differentiation in Rhabdomyosarcoma. Cancers. 2023 May 18;15(10):2823.

354. Zhang TY, Meaney MJ. Epigenetics and the environmental regulation of the genome and its function. Annual Review of Psychology. 2010 Jan 10;61:439-66.

355. Poganik JR, Zhang B, Baht G, Kerepesi C, Yim SH, Lu A, Haghani A, Gong T, Hedman A, Andolf E, Pershagen G. Biological age is increased by stress and restored upon recovery. bioRxiv. 2022:2022-05.

356. Chakravarti D, LaBella KA, DePinho RA. Telomeres: history, health, and hallmarks of aging. Cell. 2021 Jan 21;184(2):306-22.

357. Lustig A, Liu HB, Metter EJ, An Y, Swaby MA, Elango P, Ferrucci L, Hodes RJ, Weng NP. Telomere shortening, inflammatory cytokines, and anti-cytomegalovirus antibody follow distinct age-associated trajectories in humans. Frontiers in Immunology. 2017 Aug 24;8:1027.

358. Rea IM, Gibson DS, McGilligan V, McNerlan SE, Alexander HD, Ross OA. Age and age-related diseases: role of inflammation triggers and cytokines. Frontiers in Immunology. 2018:586.

359. Wagner W. The link between epigenetic clocks for aging and senescence. Frontiers in Genetics. 2019 Apr 3;10:303.

360. Schrempft S, Belsky DW, Draganski B, Kliegel M, Vollenweider P, Marques-Vidal P, Preisig M, Stringhini S. Associations between life-course socioeconomic conditions and the pace of aging. The Journals of Gerontology: Series A. 2022 Nov;77(11):2257-64.

361. Simpson DJ, Chandra T. Epigenetic age prediction. Aging cell. 2021 Sep;20(9):e13452.

362. Tuttle CS, Waaijer ME, Slee-Valentijn MS, Stijnen T, Westendorp R, Maier AB. Cellular senescence and chronological age in various human tissues: A systematic review and meta-analysis. Aging Cell. 2020 Feb;19(2):e13083.

363. Firsanov D, Zacher M, Tian X, Zhao Y, George JC, Sformo TL, Tombline G, Biashad SA, Gilman A, Hamilton N, Patel A. DNA repair and anti-cancer mechanisms in the longest-living mammal: the bowhead whale. bioRxiv. 2023:2023-05.

364. Kale A, Sharma A, Stolzing A, Desprez PY, Campisi J. Role of immune cells in the removal of deleterious senescent cells. Immunity & Ageing. 2020 Dec;17(1):1-9.

365. von Kobbe C. Targeting senescent cells: approaches, opportunities, challenges. Aging. 2019;11(24):12844-12861.

366. Da Silva R, Conde D, Baudish A, Colchero F. Slow and negligible senescence among testudines challenges evolutionary theories of senescence. Science. 2022 Jun; Vol 376, Issue 6600: 1466-1470.

367. Quesada V, Freitas-Rodríguez S, Miller J, Pérez-Silva JG, Jiang ZF, Tapia W, Santiago-Fernández O, Campos-Iglesias D, Kuderna LF, Quinzin M, Álvarez MG. Giant tortoise genomes provide insights into longevity and age-related disease. Nature ecology & evolution. 2019 Jan;3(1):87-95.

368. Horn AJ, Carter CS. Love and longevity: A Social Dependency Hypothesis. Comprehensive Psychoneuroendocrinology. 2021 Nov 1;8:100088.

369. Taylor, L. What Causes Aging? An Epigenetics Study From The Sinclair Lab May Have an Answer. Bowdoin Science Journal. 2023 ARP 2.

370. Yang JH, Petty CA, Dixon-McDougall T, Lopez MV, Tyshkovskiy A, Maybury-Lewis S, Tian X, Ibrahim N, Chen Z, Griffin PT, Arnold M. Chemically induced reprogramming to reverse cellular aging. Aging. 2023 Jul 12;15.

371. Garo-Pascual M, Gaser C, Zhang L, Tohka J, Medina M, Strange BA. Brain structure and phenotypic profile of superagers compared with age-matched older adults: a longitudinal analysis from the Vallecas Project. The Lancet Healthy Longevity. 2023 Jul 12.

372. Zhou X, Yuan W, Xiong X, Zhang Z, Liu J, Zheng Y, Wang J, Liu

J. HO-1 in bone biology: Potential therapeutic strategies for osteoporosis. Frontiers in Cell and Developmental Biology. 2021:3405.

373. Molnar V, Matišić V, Kodvanj I, Bjelica R, Jeleč Ž, Hudetz D, Rod E, Čukelj F, Vrdoljak T, Vidović D, Starešinić M. Cytokines and chemokines involved in osteoarthritis pathogenesis. International Journal of Molecular Sciences. 2021 Aug 26;22(17):9208.

374. Qi W, Mei Z, Sun Z, Lin C, Lin J, Li J, Ji JS, Zheng Y. Exposure to multiple air pollutants and the risk of fractures: a large prospective population-based study. Journal of Bone and Mineral Research. 2023 Jun 21.

375. O'Mahony LF, Srivastava A, Mehta P, Ciurtin C. Is fibromyalgia associated with a unique cytokine profile? A systematic review and meta-analysis. Rheumatology. 2021 Jun;60(6):2602-14.

376. Caxaria S, Bharde S, Fuller AM, Evans R, Thomas B, Celik P, Dell'Accio F, Yona S, Gilroy D, Voisin MB, Wood JN. Neutrophils infiltrate sensory ganglia and mediate chronic widespread pain in fibromyalgia. Proceedings of the National Academy of Sciences. 2023 Apr 25;120(17):e2211631120.

377. Docherty S, Harley R, McAuley JJ, Crowe LA, Pedret C, Kirwan PD, Siebert S, Millar NL. The effect of exercise on cytokines: implications for musculoskeletal health: a narrative review. BMC Sports Science, Medicine and Rehabilitation. 2022 Dec;14(1):1-4.

378. Gevirtz R. Distinguishing sources of pain: Central vs peripheral mediation. Pelviperineology. 2020 Mar 1;39(1):13-8.

379. Meiring AR, de Kater EP, Stadhouder A, van Royen BJ, Breedveld P, Smit TH. Current models to understand the onset and progression of scoliotic deformities in adolescent idiopathic scoliosis: a systematic review. Spine Deformity. 2023 May;11(3):545-58.

380. Deyo RA, Mirza SK, Turner JA, Martin BI. Overtreating chronic back pain: time to back off? The Journal of the American Board of Family Medicine. 2009 Jan 1;22(1):62-8.

381. Buchbinder R, van Tulder M, Öberg B, Costa LM, Woolf A, Schoene M, Croft P, Hartvigsen J, Cherkin D, Foster NE, Maher CG. Low back pain: a call for action. The Lancet. 2018 Jun 9;391(10137):2384-8.

382. Deyo RA, Nachemson A, Mirza SK. Spinal-fusion surgery—the case for restraint. The Spine Journal. 2004 Sep 10;4(5):S138-42.

383. Weston EB, Hassett AL, Khan SN, Weaver TE, Marras WS. Cognitive dissonance increases spine loading in the neck and low back. Ergonomics. 2023 Mar 9:1-5.

384. Ranney RM, Maguen S, Bernhard PA, Holder N, Vogt D, Blosnich JR, Schneiderman AI. Moral injury and chronic pain in veterans. Journal of Psychiatric Research. 2022 Nov 1;155:104-11.

385. Bland J, Haase, D, Dwarka, V, Smith. R. Epigenetic Age Testing and the Future Landscape of Epigenetic Diagnostics with subject-matter experts and thought leaders. TruDiagnostic 2023.

386. Kheirandish-Gozal L, Gozal D. Obstructive sleep apnea and inflammation: proof of concept based on two illustrative cytokines. International Journal of Molecular Sciences. 2019 Jan 22;20(3):459.

387. Piber D, Cho JH, Lee O, Lamkin DM, Olmstead R, Irwin MR. Sleep disturbance and activation of cellular and transcriptional mechanisms of inflammation in older adults. Brain, Behavior, and Immunity. 2022 Nov 1;106:67-75.

388. Aghelan Z, Karima S, Khazaie H, Abtahi SH, Farokhi AR, Rostampour M, Bahrehmand F, Khodarahmi R. IL-1a and TNF-a as an inducer for ROS-mediated NLRP1/NLRP3 inflammasomes activation in mononuclear blood cells from individuals with chronic insomnia disorder. European Journal of Neurology. 2022 Sep 1.

389. Kuna K, Szewczyk K, Gabryelska A, Białasiewicz P, Ditmer M, Strzelecki D, Sochal M. Potential Role of Sleep Deficiency in Inducing Immune Dysfunction. Biomedicines. 2022 Sep 1;10(9):2159.

390. Costa G. The impact of shift and night work on health. Applied Ergonomics. 1996 Feb 1;27(1):9-16.

391. Muscogiuri G, Poggiogalle E, Barrea L, Tarsitano MG, Garifalos F, Liccardi A, Pugliese G, Savastano S, Colao A, Alviggi C, Aprano S. Exposure to artificial light at night: A common link for obesity and cancer? European Journal of Cancer. 2022 Sep 1;173:263-75.

392. Balmes JR. Can Breathing Poor-quality Air Lead to Poor-quality Sleep in Children?. American Journal of Respiratory and Critical Care Medicine. 2023 Mar 1;207(5):510-1.

393. Rosada C, Bauer M, Golde S, Metz S, Roepke S, Otte C, Buss C, Wingenfeld K. Childhood trauma and cortical thickness in healthy women, women with post-traumatic stress disorder, and women with borderline personality disorder. Psychoneuroendocrinology. 2023 Jul 1;153:106118.

394. Nash K, Minhas S, Metheny N, Gokhale KM, Taylor J, Bradbury-Jones C, Bandyopadhyay S, Nirantharakumar K, Adderley NJ, Chandan JS. Exposure to Domestic Abuse and the Subsequent Development of Atopic Disease in Women. The Journal of Allergy and Clinical Immunology: In Practice. 2023 May 5.

395. Guo Y, Bian J, Chen Z, Fishe JN, Zhang D, Braithwaite D, George TJ, Shenkman EA, Licht JD. Cancer incidence after asthma diagnosis: Evidence from a large clinical research network in the United States. Cancer Medicine. 2023 Mar 31.

396. Trasande L, Nelson ME, Alshawabkeh A, Barrett ES, Buckley JP, Dabelea D, Dunlop AL, Herbstman JB, Meeker JD, Naidu M, Newschaffer C. Prenatal phthalate exposure and adverse birth outcomes in the U.S.A.: a prospective analysis of births and estimates of attributable burden and costs. The Lancet Planetary Health. 2024 Feb 1;8(2):e74-85.

397. Tanya Nagahawatte N, Goldenberg RL. Poverty, maternal health, and adverse pregnancy outcomes. Annals of the New York Academy of Sciences. 2008 Jun;1136(1):80-5.

398. Hardeman, R.R., Kheyfets, A., Mantha, A.B. et al. Developing Tools to Report Racism in Maternal Health for the CDC Maternal Mortality Review

Information Application (MMRIA): Findings from the MMRIA Racism & Discrimination Working Group. Matern Child Health J 26, 661–669 (2022).

399. Salazar EG, Montoya-Williams D, Passarella M, McGann C, Paul K, Murosko D, Peña MM, Ortiz R, Burris HH, Lorch SA, Handley SC. County-Level Maternal Vulnerability and Preterm Birth in the U.S.. JAMA Network Open. 2023 May 1;6(5):e2315306-.

400. Ashorn P, Ashorn U, Muthiani Y, Aboubaker S, Askari S, Bahl R, Black RE, Dalmiya N, Duggan CP, Hofmeyr GJ, Kennedy SH. Small vulnerable newborns—big potential for impact. The Lancet. 2023 May 5.

401. McNestry C, Killeen SL, Crowley RK, McAuliffe FM. Pregnancy complications and later life women's health. Acta Obstetricia et Gynecologica Scandinavica. 2023 May;102(5):523-31.

402. Malutan AM, Costin N, Ciortea R, Dan MI. Variation of Anti-inflammatory Cytokines in Relationship with Menopause. Applied Medical Informatics. 2013 Jun 6;32(2):30-8.

403. Stoner R, Camilleri V, Calleja-Agius J, Schembri-Wismayer P. The cytokine-hormone axis–the link between premenstrual syndrome and postpartum depression. Gynecological Endocrinology. 2017 Aug 3;33(8):588-92.

404. Udenze I, Amadi C, Awolola N, Makwe CC. The role of cytokines as inflammatory mediators in preeclampsia. Pan African Medical Journal. 2015;20(1).

405. Rhee KY, Goetzl L, Unal R, Cierny J, Flood P. The relationship between plasma inflammatory cytokines and labor pain. Anesthesia & Analgesia. 2015 Sep 1;121(3):748-51.

406. Guerinot GT, Gitomer SD, Sanko SR. Postpartum patient with toxic shock syndrome. Obstetrics and Gynecology. 1982 Jun 1;59(6 Suppl):43S-6S.

407. Aschbacher K, Hagan M, Steine IM, Rivera L, Cole S, Baccarella A, Epel ES, Lieberman A, Bush NR. Adversity in early life and pregnancy are immunologically distinct from total life adversity: macrophage-associated phenotypes in women exposed to interpersonal violence. Translational Psychiatry. 2021 Jul 20;11(1):1-9.

408. Kobayashi H, Yamada Y, Morioka S, Niiro E, Shigemitsu A, Ito F. Mechanism of pain generation for endometriosis-associated pelvic pain. Archives of Gynecology and Obstetrics. 2014 Jan;289(1):13-21.

409. Ebejer K, Calleja-Agius J. The role of cytokines in polycystic ovarian syndrome. Gynecological Endocrinology. 2013 Jun 1;29(6):536-40.

410. Jiang YH, Jhang JF, Hsu YH, Ho HC, Wu YH, Kuo HC. Urine cytokines as biomarkers for diagnosing interstitial cystitis/bladder pain syndrome and mapping its clinical characteristics. American Journal of Physiology-Renal Physiology. 2020 Jun 1;318(6):F1391-9.

411. Solmi M, Veronese N, Favaro A, Santonastaso P, Manzato E, Sergi G, Correll CU. Inflammatory cytokines and anorexia nervosa: A meta-analysis of cross-sectional and longitudinal studies. Psychoneuroendocrinology. 2015 Jan 1;51:237-52.

412. Walton E, Bernardoni F, Batury VL, Bahnsen K, Larivière S, Abbate-Daga G, Andres-Perpiña S, Bang L, Bischoff-Grethe A, Brooks SJ, Campbell IC. Brain Structure in Acutely Underweight and Partially Weight-Restored Individuals with Anorexia Nervosa-A Coordinated Analysis by the ENIGMA Eating Disorders Working Group. Biological Psychiatry. 2022 May 31.

413. Sanbe A, Tanaka Y, Fujiwara Y, Tsumura H, Yamauchi J, Cotecchia S, Koike K, Tsujimoto G, Tanoue A. a1 Adrenoceptors are required for normal male sexual function. British journal of pharmacology. 2007 Oct;152(3):332-40.

414. Potegal M, Nordman JC. Non-angry aggressive arousal and angriffsberietschaft: a narrative review of the phenomenology and physiology of proactive/offensive aggression motivation and escalation in people and other animals. Neuroscience & Biobehavioral Reviews. 2023 Feb 21:105110.

415. Kuo CT, Kawachi I. County-Level Income Inequality, Social Mobility, and Deaths of Despair in the U.S., 2000-2019. JAMA Network Open. 2023 Jul 3;6(7):e2323030-.

416. Fredriksen-Goldsen KI, Simoni JM, Kim HJ, Lehavot K, Walters KL, Yang J, Hoy-Ellis CP, Muraco A. The health equity promotion model: Reconceptualization of lesbian, gay, bisexual, and transgender (LGBT) health disparities. American Journal of Orthopsychiatry. 2014 Nov;84(6):653.

417. Tordoff DM, Wanta JW, Collin A, Stepney C, Inwards-Breland DJ, Ahrens K. Mental health outcomes in transgender and nonbinary youths receiving gender-affirming care. JAMA Network Open. 2022 Feb 1;5(2):e220978-.

418. Pääbo S. The mosaic that is our genome. Nature. 2003 Jan;421(6921):409-12.

419. Botwinick L, Reid A, Wyatt R, Whittington J. Addressing institutional racism in healthcare organizations. Healthcare Executive. 2021 May;36(3):42-43.

420. Ellingsen DM, Isenburg K, Jung C, Lee J, Gerber J, Mawla I, Sclocco R, Grahl A, Anzolin A, Edwards RR, Kelley JM. Brain-to-brain mechanisms underlying pain empathy and social modulation of pain in the patient-clinician interaction. Proceedings of the National Academy of Sciences. 2023 Jun 27;120(26):e2212910120

421. Hojat M, Vergare MJ, Maxwell K, Brainard G, Herrine SK, Isenberg GA, Veloski J, Gonnella JS. The devil is in the third year: a longitudinal study of erosion of empathy in medical school. Academic Medicine. 2009 Sep 1;84(9):1182-91.

422. Wilkinson H, Whittington R, Perry L, Eames C. Examining the relationship between burnout and empathy in healthcare professionals: A systematic review. Burnout Research. 2017 Sep 1;6:18-29.

423. Klopack ET, Crimmins EM, Cole SW, Seeman TE, Carroll JE. Social stressors associated with age-related T lymphocyte percentages in older U.S. adults: Evidence from the Health and Retirement Study. medRxiv. 2022

424. National Center for Health Statistics – U.S. Health, United States, 2015: With special feature on racial and ethnic health disparities. 2016.

425. Van der Kolk BA. The body keeps the score: Memory and the evolving psychobiology of posttraumatic stress. Harvard Review of Psychiatry. 1994 Jan 1;1(5):253-65.

426. Bundy JD, Mills KT, He H, LaVeist TA, Ferdinand KC, Chen J, He J. Social determinants of health and premature death among adults in the U.S.A. from 1999 to 2018: a national cohort study. The Lancet Public Health. 2023 Jun 1;8(6):e422-31

427. Johnson CO, Boon-Dooley AS, DeCleene NK, Henny KF, Blacker BF, Anderson JA, Afshin A, Aravkin A, Cunningham MW, Dieleman JL, Feldman RG. Life expectancy for White, Black, and Hispanic race/ethnicity in U.S. states: trends and disparities, 1990 to 2019. Annals of Internal Medicine. 2022 Aug;175(8):1057-64.

428. Dumornay NM, Lebois LA, Ressler KJ, Harnett NG. Racial disparities in adversity during childhood and the false appearance of race-related differences in brain structure. American Journal of Psychiatry. 2023 Feb 1;180(2):127-38.

429. Hanscom D, Clawson DR, Porges SW, Bunnage R, Aria L, Lederman S, Taylor J, Carter CS. Polyvagal and global cytokine theory of safety and threat Covid-19–plan B. SciMedicine Journal. 2020 Aug 8;2:9-27.

430. Wang S, Quan L, Chavarro JE, Slopen N, Kubzansky LD, Koenen KC, Kang JH, Weisskopf MG, Branch-Elliman W, Roberts AL. Associations of Depression, Anxiety, Worry, Perceived Stress, and Loneliness Prior to Infection With Risk of Post-COVID-19 Conditions. JAMA Psychiatry. 2022 Nov.

431. Phetsouphanh C, Darley DR, Wilson DB, Howe A, Munier C, Patel SK, Juno JA, Burrell LM, Kent SJ, Dore GJ, Kelleher AD. Immunological dysfunction persists for 8 months following initial mild-to-moderate SARS-CoV-2 infection. Nature Immunology. 2022 Feb;23(2):210-6.

432. Jiang Y, Rubin L, Peng T, Liu L, Xing X, Lazarovici P, Zheng W. Cytokine storm in COVID-19: from viral infection to immune responses, diagnosis and therapy. International Journal of Biological Sciences. 2022;18(2):459.

433. SJ, Dornbush R, Lynch S, Shahar S, Klepacz L, Karmen CL, Chen D, Lobo SA, Lerman D. Neuropsychological, Medical, and Psychiatric Findings After Recovery from Acute COVID-19: A Cross-sectional Study. Journal of the Academy of Consultation-liaison Psychiatry. 2022 Jan 25.

434. Lee MH, Perl DP, Steiner J, Pasternack N, Li W, Maric D, Safavi F, Horkayne-Szakaly I, Jones R, Stram MN, Moncur JT. Neurovascular injury with complement activation and inflammation in COVID-19. Brain. 2022 Jul;145(7):2555-68.

435. White-Dzuro G, Gibson LE, Zazzeron L, White-Dzuro C, Sullivan Z, Diiorio DA, Low SA, Chang MG, Bittner EA. Multisystem effects of COVID-19: a concise review for practitioners. Postgraduate Medicine. 2021 Jan 2;133(1):20-7.

436. Shabnam S, Razieh C, Dambha-Miller H, Yates T, Gillies C, Chudasama YV, Pareek M, Banerjee A, Kawachi I, Lacey B, Morris EJ. Socioeconomic inequalities of Long COVID: a retrospective population-based cohort study in the United Kingdom. Journal of the Royal Society of Medicine. 2023 May 10:01410768231168377.

437. Attia, P. Twenty Years of the Human Genome: Growing Older and Wiser. 2023, Apr 8.

438. Hsieh, P. Does Your Doctor Work for a Private Equity Company – and Should You Care? Forbes. Jan 31, 2022.

439. Gray MW. Mitochondrial evolution. Cold Spring Harbor Perspectives in Biology. 2012 Sep 1;4(9):a011403 icard.

440. M, Sandi C. The social nature of mitochondria: Implications for human health. Neuroscience & Biobehavioral Reviews. 2021 Jan 1;120:595-610.

441. Lucero-Prisno III DE, Shomuyiwa DO, Kouwenhoven MB, Dorji T, Odey GO, Miranda AV, Ogunkola IO, Adebisi YA, Huang J, Xu L, Obnial JC. Top 10 public health challenges to track in 2023: Shifting focus beyond a global pandemic. Public Health Challenges. 2023 Jun;2(2):e86.

442. Bezruchka S. Inequality Kills Us All: COVID-19's Health Lessons for the World. Taylor & Francis; 2022 Nov 28.

443. Willroth EC, Young G, Tamir M, Mauss IB. Judging emotions as good or bad: Individual differences and associations with psychological health. Emotion. Epub ahead of print. PMID: 36913276. 2023 Mar 13.

444. Masedo AI, Rosa Esteve M. Effects of suppression, acceptance and spontaneous coping on pain tolerance, pain intensity and distress. Behav Res Ther. 2007.

445. Pennebaker JW, Chung CK. 18 Expressive Writing: Connections Physical and Mental Health to. The Oxford handbook of health psychology. 2014:417.

446. Staples JK, Gordon JS, Hamilton M, Uddo M. Mind-body skills groups for treatment of war-traumatized veterans: A randomized controlled study. Psychological Trauma: Theory, Research, Practice, and Policy. 2020 Mar.

447. Calabrese, L. Humanities and rheumatology: The time is now (again) Healio January 23, 2023.

448. Where the Arts Meet Health, Aspen Ideas (blog). 2023, Apr,11.

449. Takahashi A, Flanigan ME, McEwen BS, Russo SJ. Aggression, social stress, and the immune system in humans and animal models. Frontiers in Behavioral Neuroscience. 2018 Mar 22;12:56.

450. Hirabayashi N, Honda T, Hata J, Furuta Y, Shibata M, Ohara T, Tatewaki Y, Taki Y, Nakaji S, Maeda T, Ono K. Association Between Frequency of Social Contact and Brain Atrophy in Community-Dwelling Older People Without Dementia: The JPSC-AD Study. Neurology. 2023 Jul 12.

451. Vivek Murthy.The New York Times. How to Feel Less Lonely, According to the Surgeon General. 2023, May 5.

452. Yin J, Pan Y, Zhang Y, Hu Y, Luo J. Distinct inter-brain synchronization patterns during group creativity under threats in cooperative and competitive contexts. Thinking Skills and Creativity. 2023 Jul 10:101366.

453. Guo Q, Li S, Liang J, Yu X, Lv Y. Mindfulness may be associated with less prosocial engagement among high intelligence individuals. Scientific Reports. 2023 Mar 14;13(1):4208.

454. Folk D, Dunn E. A systematic review of the strength of evidence for the most commonly recommended happiness strategies in mainstream media. Nature Human Behaviour. 2023 Jul 20:1-1.

455. Daly M. Stockton Economic Empowerment Demonstration: A Case Study of Basic Income. University of California, Irvine; 2022.

456. Joseph R. The Split Brain: Two Brains-Two Minds. Consciousness and the Universe. 2011:108-52.

457. Galbraith ED, Barrington-Leigh C, Miñarro S, Álvarez-Fernández S, Attoh EM, Benyei P, Calvet-Mir L, Carmona R, Chakauya R, Chen Z, Chengula F. High life satisfaction reported among small-scale societies with low incomes. Proceedings of the National Academy of Sciences. 2024 Feb 13;121(7):e2311703121.

458. Haas S, How to Be Happy – Public policies based on culture are making people happy. Psychology Today Posted, 2022. June 25.

459. Colston, P. The Finnish Secret to Happiness? Knowing When You Have Enough. New York Times. 2023, Apr, 1.

460. Cole LE. Japanese Culture-Bound Disorders: The Relationship between Taijin Kyofusho, Hikikomori, and Shame. Online Submission. 2013 Aug.

461. Braghieri L, Levy RE, Makarin A. Social media and mental health. American Economic Review. 2022 Nov 1;112(11):3660-93.

462. Stamkou E, Brummelman E, Dunham R, Nikolic M, Keltner D. Awe sparks prosociality in children. Psychological science. 2023 Apr;34(4):455-67

463. Ghebreyesus TA. Achieving health for all requires action on the economic and commercial determinants of health. The Lancet. 2023 Apr 8;401(10383):1137-9.

464. Bor J, Stokes AC, Raifman J, Venkataramani A, Bassett MT, Himmelstein D, Woolhandler S. Missing Americans: Early death in the United States—1933–2021. PNAS nexus. 2023 Jun;2(6):pgad173.

465. Samarasekera DD, Lee SS, Yeo JH, Yeo SP, Ponnamperuma G. Empathy in health professions education: What works, gaps and areas for improvement. Medical Education. 2023 Jan;57(1):86-101.

466. Kiehl JT. Climate Chaos: A Complex Issue. Psychological Perspectives. 2023 Jan 2;66(1):45-56.

467. Cianconi P, Hanife B, Hirsch D, Janiri L. Is climate change affecting mental health of urban populations?. Current Opinion in Psychiatry. 2023 May 1;36(3):213-8.

Both threat and safety compound – lean into safety

Disclosures

TVST is a proof-of-concept collection of ideations, constructs, and knowledge put into a long narrative. Yet, the wisdom of TVST will not be found within these functions of the "sapiocortex," but by separating from the "sapiocortex" and into the application of The Big Bang Principles of TVST.

Ali F Oeus = All of Us